Sustainable Product Design and Development

Industrial Engineering: Challenges in Production and Service Industries

Series Editor: Anil Mital

This new book series is intended to cover a wide range of subject matter in both production and service industries from an Industrial Engineering perspective. The word production, in this context, may be considered a synonym of the word manufacturing; production is, however, much broader and global, applicable to manufacturing industries as well as to process industries such as petrochemical, mining and pharmaceutical. Service sector industries, on the other hand, refer to industries such as transportation, health, energy and banking. Other book series by other publishers, such as Prentice-Hall and Elsevier in Industrial Engineering, have focused on very traditional areas: engineering economy and operations research. This new book series will focus instead on topical application areas such as information processing into the healthcare industry, manufacturing economics and additive manufacturing. The new book series will include books that target the industrial practitioner as well as the reference user and will emphasize both the "how to" type of books as well as books that may be used as texts or references. The series may also include encyclopedias. Examples of targeted titles are: Green Design and Sustainability, Production Economics, Information Processing and Display in the Health Industry, Energy Audit in Industry, Additive Manufacturing, to name just a few…

Production Economics
Evaluating Costs of Operations in Manufacturing and Service Industries
Anoop Desai and Aashi Mital

Sustainable Product Design and Development
Anoop Desai and Anil Mital

For more information on this series, please visit: www.routledge.com/Production-Economics-Evaluating-Costs-of-Operations-in-Manufacturing-and/Desai-Mital/p/book/9781138033269

Sustainable Product Design and Development

Anoop Desai and Anil Mital

CRC Press
Taylor & Francis Group
Boca Raton London New York

CRC Press is an imprint of the
Taylor & Francis Group, an informa business

First edition published 2021
by CRC Press
6000 Broken Sound Parkway NW, Suite 300
Boca Raton, FL 33487-2742

and by CRC Press
2 Park Square, Milton Park, Abingdon, Oxon OX14 4RN

Library of Congress Cataloging-in-Publication Data
Names: Desai, Anoop, author. | Mital, Anil, author.
Title: Sustainable product design and development/Anoop Desai and Anil Mital.
Description: First edition. | Boca Raton: CRC Press, 2021. |
Includes bibliographical references and index.
Identifiers: LCCN 2020027792 (print) | LCCN 2020027793 (ebook) |
ISBN 9780367343217 (hardback) | ISBN 9780429327803 (ebook)
Subjects: LCSH: Product design–Environmental aspects. | Sustainable design.
Classification: LCC TS171.4 .D468 2021 (print) |
LCC TS171.4 (ebook) | DDC 628–dc23
LC record available at https://lccn.loc.gov/2020027792
LC ebook record available at https://lccn.loc.gov/2020027793

ISBN: 978-0-367-34321-7 (hbk)
ISBN: 978-0-429-32780-3 (ebk)

To our family and colleagues

Contents

Preface

Conventional product design is based on the cradle-to-grave manufacturing system also referred to as a linear manufacturing system. It builds upon familiar topics such as design for manufacturing, consideration of traditional manufacturing processes and materials and so on. In such a system, customer requirements are converted into engineering specifications which are in turn embodied into the actual product. This familiar practice has been widely resorted to through most of the twentieth century. We have presented a comprehensive overview of this process in a previous book entitled '*Product Development: A Structured approach to consumer product development, design and manufacture* (Mital et al., 2014. Second Edition. Elsevier).' The traditional approach to product design focusing on functionality, usability, design review, consideration of materials and manufacturing processes etc has been elaborated upon at length in that book. However, times are changing rapidly.

Increased environmental awareness has prompted stricter government legislation in terms of product take back. Such legislation requires companies to reduce, reuse and recycle. There is a demand for companies to design components and products that would result in minimal environmental impact. This dire situation has prompted at least two dozen US states to pass laws concerning environment-friendly disposal of electronic waste, for example.

For instance, the electronics waste recycling act of 2003 passed by California limits hazardous substances used in certain electronic products sold in that state. In addition, an electronic waste recycling fee is also collected at the point of sale. As far as the collector and recycler are concerned, it requires the disbursement of recovery and recycling payments to certain qualified entities to cover the average cost of electronic waste collection as well as recycling (Electronic Waste Recycling Act, 2003). The Covered Device Recycling Act of 2010 passed by the state of Pennsylvania started the landfill ban in January 2013. According to this law, manufacturers are responsible for proper disposal of electronics. Also, they should register with the state to sell their products. They should establish and conduct ongoing recycling programs (CDRA, Act 108, 2010). The Responsible Electronics Recycling Act (RERA) was introduced in 2013 with the intent to promote investment in developing domestic infrastructure in order to manage the electronic waste generated domestically by the US. The act was designed to create tens of thousands of good jobs, expand trade and in the process achieve economic strength (RERA, 2013). In 2012, the Vermont legislature unanimously passed the Universal Recycling and Composting Law that offers the residents of the state a new set of systems and tools to keep as much as possible out of the landfills. According to this law, landfill bans will be phased in through the year 2020. The law also introduced as "pay as you throw" variable pricing rate in order to incentivize trash reduction (Universal Recycling and Composting Law, 2012).

The necessity to not only design products for functionality, but for sustainability as well is now self-evident. The stage has been set for a complete reconsideration of the product design process. In order to use natural resources wisely, it is essential to consider end-of-life (EoL) objectives such as reuse, recycling and remanufacturing during the product design stage itself instead of as an afterthought.

This book has been written to enable sustainable product design and development. To that end, it considers prolonging the life of a product. It is well known that every product will eventually fail. However, in order to minimize its environmental impact, it is necessary to extend the lifespan of a product. The book will consider issues such as design for disassembly in depth as well as design for assembly and design for maintenance. The chapter on design for disassembly will deal with partial and complete disassembly. It will also present concepts on disassembly sequence plans (DSPs) as well as intelligent systems that will enable the designer to introduce disassembly as one of the integral Xs of the DfX algorithm. Within the realm of product assembly, issues such as modularization for sustainable product development will be addressed. The chapter on reuse, recycling and remanufacturing will present design guidelines to design products for those specific end-of-life options (such as reuse, recycling and remanufacturing). Costing issues pertaining to sustainable product design are also addressed. Fundamental concepts such as consideration of materials and processes will be embedded throughout the chapters. Some industrial case studies will also be included to help the practitioner apply principles presented in the relevant chapters in the book.

Anoop Desai
Anil Mital

Author Biographies

Anoop Desai is associate professor of mechanical engineering at Georgia Southern University. He received his PhD in industrial engineering from the University of Cincinnati in 2006. His areas of research interest include product development, design for manufacturing, six sigma and quality control and engineering economic analysis.

Dr. Desai is the author of *Production Economics: Evaluating Costs of Operations in Manufacturing and Service Industries*, *Product Development: A Structured Approach to Consumer Product Design, Development and Manufacture*, *Fundamentals of Work Measurement: What Every Engineer Should Know* and *Global Sustainability and Innovation* (Ed.).

Dr. Desai is also the author of 30 journal papers mainly focusing on the area of Design for X, environment conscious design and manufacturing and the economics of production processes. He is Elsevier Science's one of the most cited authors.

Anil Mital is a retired professor of manufacturing design and engineering at the University of Cincinnati. He is also former professor and director of industrial engineering and professor of physical medicine and rehabilitation at the University of Cincinnati. Dr. Mital is the founding Editor-in-Chief Emeritus of Elsevier's *International Journal of Industrial Ergonomics* and is the founding Editor-in-Chief Emeritus of the *International Journal of Industrial Engineering—Theory, Applications, and Practice*. Dr. Mital has authored and coauthored nearly 500 publications, including 200 journal articles and 23 books. He has made over 200 technical presentations in various parts of the world. He frequently conducted seminars in different countries on a wide range of topics, such as work design, engineering economy, facilities planning, human-centered manufacturing, ergonomics and product design. Dr. Mital is Fellow of the Institute of Industrial Engineers (IIE) and the Human Factors and Ergonomics Society (HFES). He also is a recipient of IIE's David F. Baker Distinguished Research Award, HFES's Paul M. Fitts Educational Award and the Society of Automotive Engineers' Ralph Teetor Educational Award. Dr. Mital has been recognized by the Engineering Economy Division of IIE through its Eugene Grant Award and by the Society of Work Sciences through its M.M. Ayoub Award.

1 Introduction

1.1 DEFINITION OF SUSTAINABILITY

In 1987, The Brundtland commission defined sustainability as "Meeting the needs of the present without compromising the ability of future generations to meet their own needs (Fiksel et al., 1998)." The gist of sustainability is that all commercial activity, including product development, should preserve the quality of ecological and societal environment for future generations. Most commercial activity leads to consumption of natural resources. That is what enables us to meet our needs in the present. But natural resources have been dwindling for a while and environmental pollution has seen a drastic and meteoric rise. The situation is beginning to get dire, so much so that it has attracted the attention of lawmakers all around the world. It has prompted them to pass restrictive environmental legislation that puts the onus upon the manufacturer to address and curb the environmental impact of their products.

In short, sustainability tries to address two mutually conflicting goals: consuming natural resources in the present as well as conserving them for the future. The conflict can be stated thus: "How can you consume and conserve simultaneously?" And that is the dilemma.

In order to find the answer, the problem needs to be solved creatively. It is imperative that the process of sustainable product design constitutes paradigms that differ from conventional counterpart. Products will need to be designed in order to conserve material and energy. This can be achieved by designing products so that they are easy to reuse, recycle and remanufacture. This goal can in turn be met by incorporating design practices such as design for assembly, disassembly and maintenance into the product design process. These goals and design practices will be presented in detail in future chapters of this book.

Sustainability of a product is evaluated using the so-called triple bottom line also referred to as the three pillars of sustainability, namely: economic, environmental and social. Each of the three pillars needs to be assessed through quantitative criteria. Doing so enables the designer to truly comprehend to what extent their design solutions have been effective. We shall address each pillar in some detail as follows.

1.1.1 Performance Evaluation of the Economic Pillar

Traditional metrics of economic performance include cash flow and profitability. However, for a commercial activity to be truly sustainable, this is not enough. As we have said earlier, the answer lies in solving the problem creatively. It is important to go over and beyond the call of duty. The following issues need to be addressed for economic sustainability (Fiksel et al., 1998):

- The utilization of material, energy, capital and human resources costs money. Often, it is difficult to foresee this cost. As a result, it is also referred to as hidden cost. This cost should be quantified.
- Industrial production and consumption impact the environment. This is referred to as external impact. The uncertain future costs associated with it should be estimated.
- It is also necessary to estimate life cycle costs and benefits incurred during the lifecycle of a product. These costs and benefits are incurred by a host of stakeholders such as customers, employees, communities and so on across a product or process lifecycle.

Industry is at the forefront of trying to solve the problem. Take the case of Chrysler Corporation, for example. During the process of designing several new automotive components, Chrysler took into account the direct cost associated with each design option. This cost was in fact a potentially hidden and contingent cost. The potentially hidden costs were evaluated using activity-based costing methods. The company developed proprietary risk factors to account for the contingent costs. For example, when developing an oil filter, they considered the possible liabilities associated with disposing used oil (Armstrong and White, 1997). It was realized that the design option with the lowest material and labor costs did not simultaneously have the lowest life cycle cost. This was due to the liability cost, which was classified as a hidden cost.

1.1.2 Performance Evaluation of the Environmental Pillar

The environmental pillar of sustainability primarily addresses the problem of resource consumption, or rather exploitation of natural resources. As far as consumption of natural resources is concerned, a product would qualify as a sustainable product if it achieves the following objectives:

- Maximizes value in the "triple bottom line" sense.
- Simultaneously minimizes resource consumption.

But what is a resource? How do you define it? As far as sustainability is concerned, resources are defined broadly. A resource can be defined as a natural or anthropogenic (manmade) stock that is necessary in order to create, use and dispose of a product (Fiksel et al., 1998). Material (such as iron, steel, glass, copper, wood, etc.), energy, land and labor are all examples of a resource, and there are many such resources.

The stakeholders of a company benefit following the "value" that is attributable to the product. Thus, a valuable product could result in improved profitability, liberation of time, reduced pollution and so on.

Feldman et al. (1997) have demonstrated that there exists a plausible connection between improved environmental performance and increased shareholder value. As a result, corporations are beginning to increasingly report the environmental performance of their products and companies voluntarily (Blumberg, 1997). Environmental standards have also been developed with the objective of quantifying the impact of products on the environment

The ISO 14000 series of standards is well known in this regard. Through these standards, an international consensus was developed on the different elements of an environmental performance evaluation process (Fiksel, 1997). The relationship between environmental and economic pillars of sustainability has been assessed by numerous companies by relying upon an objective indicator, namely eco-efficiency.

Eco-efficiency is defined as a measure of environmental performance relative to economic input or output. For instance, Sony Corp utilizes an environment performance indicator (EPI) for batteries. Fiksel (1997) defines a performance indicator as a product attribute that is specific and measurable. It measures a particular aspect of sustainability. As far as Sony is concerned, the value of this indicator is computed as the economic value added over the life cycle of a product divided by the sum of nonrecyclable material consumption and amount of energy used in the production process (Lehni, 1998).

Dow Chemical, on the other hand, uses an EPI known as the Eco-Compass. It includes evaluations of such measures as mass intensity, risk potential, energy intensity, reuse, resource conservation and extent of service. Data from product lifecycle analysis is used to compute the aforementioned measures (Lehni, 1998; James, 1997).

Novo Nordisk utilizes an eco-efficiency indicator whose value is computed by dividing the indexed turnover in constant prices by the indexed resource consumption.

1.1.3 Performance Evaluation of the Societal Pillar

The societal impact of product design and corporate decisions needs to be evaluated as a component of assessing sustainability. Since the 1980s, there has been increasing media interest in the social impact of corporate operations. Policies pursued by management and the decisions made by them can result in adversely impacting publicity, damaging the brand image of a product and, finally, result in alienating customers.

Take the example of The Body Shop, which is a skin and hair care product manufacturer based in the UK. The company released its first so-called Social Report in 1995. In this report, it presented performance results on about 200 stated targets that were grouped into nine stakeholder categories. The performance results were based upon surveys and the results were presented in the form of percentages of stakeholder responses.

Similarly, an attempt was made by other companies such as British Petroleum to measure its social performance. In its social report released in 1997, the company acknowledged that creating value was important in the communities impacted by the company's operations. Doing so constituted an important measure of the company's social performance.

A standard has been proposed by the Council on Economic Priorities (CEP) to assess social accountability. This standard is called the SA 8000. It is similar to the ISO 14000 standard on environmental sustainability. It is hoped that this standard will follow in the same footsteps as its ISO 9000 and ISO 14000 counterparts. SA 8000 focuses on standardizing the evaluation of corporate commitment to human rights. It has made significant inroads into dealing with pertinent issues such as worker safety and equality (Ranganathan, 1998).

Different categories of sustainable product indicators that can be potentially quantified are presented here. It is advised that in the initial stage of the design process, designers choose the most relevant indicators and focus their design efforts on them rather than addressing all indicators simultaneously (Fiksel et al., 1998).

1.1.4 Sustainable Product Indicators for the Economic Pillar

- Direct cost:
  - i Cost of raw material.
  - ii Cost of labor.
  - iii Capital cost.
- Potentially hidden cost:
  - i Revenue obtained from recycling.
  - ii Cost of product disposal.
- Contingent cost:
  - i Cost of dealing with employee injury.
  - ii Customer warranty cost.
- Relationship cost:
  - i Cost accruing due to loss of goodwill as a result of customer concerns.
  - ii Cost accruing due to business interruptions as a result of stakeholder interventions.
- External cost:
  - i Ecosystem productivity loss.
  - ii Depletion of resources.

1.1.5 Sustainable Product Indicators for the Environmental Pillar

- Material consumption:
  - i Product and packaging mass.
  - ii Useful lifetime of a product.
  - iii Hazardous materials used in the product.
- Consumption of energy:
  - i Energy consumed over the lifecycle of the product.
  - ii Total amount of power used during operation.
- Local impact:
  - i Ease of recycling of the product (recyclability).
  - ii Impact upon local streams.

- Regional impact:
  - i Creation of smog.
  - ii Precursors to acidification of rain.
  - iii Reduction of biodiversity.
- Global impact:
  - i Carbon di oxide emissions.
  - ii Depletion of ozone.

1.1.6 Sustainable Product Indicators for the Societal Pillar

- Quality of life:
  - i Breadth of product availability.
  - ii Enhancement of knowledge or skill.
- Peace of mind:
  - i Perceived risk.
  - ii Complaints about the product from customers.
- Reduction of illness and disease:
  - i Illnesses avoided as a result of using the product.
  - ii Reduction in mortality as a result of using the product.
- Reduction in the rate of accidents and injuries:
  - i Lost time injuries.
  - ii Reportable releases.
  - iii Total number of incidents.
- Impact upon health and wellness:
  - i Provision of nutritional value.
  - ii Cost of food.

In this section of the chapter, we dealt with the basic concept of sustainability, the three pillars of sustainability and the importance of the process of sustainable product design and development. We also made a special mention of the fact this process needs to be a customized version of the generalized product design process. Thus, in order to fully grasp the nuances of sustainable product design, it is first essential to familiarize ourselves with the generalized product design process. The remainder of the chapter is dedicated to that goal. The last section of the chapter will present the ten golden rules of sustainable product design.

1.2 DEFINITION OF DESIGN AND THE ENGINEERING DESIGN PROCESS

Design is a formalized process whereby an idea or concept is transformed into tangible information. Although it is distinct from the process of making or building something, it is often confused with those terms. Caldecote (1989) has defined design as the process of converting an idea into information from which a product can be made.

Since this is a book on engineering, let us formalize the definition of engineering design. From the engineering point of view, design can be defined as the systematic application of scientific concepts, mathematical principles and creative paradigms. The goal of this application is to envision a structure, system or machine that performs an intended function.

The engineering design process utilizes six design paradigms to achieve optimum results. The design paradigms are used in the process to indicate an archetypal solution. Each paradigm is used to either describe a design solution or as an approach to design problem solving. Each of these paradigms is listed here:

1 The need for a model: Before building the final product, it is important to first make a prototype that can be tested for functionality and a myriad of other design goals. Doing so can help the designer test the concept and refine it prior to full-scale production.
2 The need for redundancy: It is important to build redundancy into product design so that any system failure does not end up being fatal. The redundancy saves the day by maintaining product functionality. It answers the question: "If the worst were to happen, what would keep the product going?"
3 The scale effect: This is an important paradigm because the scale effect can result in an inappropriate design and a failure of the prototype. The scale effect can be detrimental because reliance upon it is likely in product components being designed incorrectly.
4 Problem analysis should not be started in the middle: The problem-solving process should be started at the very beginning. Doing so enables the designer to focus on the end goal of the design process.
5 Avoid confirming a false hypothesis: Several examples can be used to validate a hypothesis. But one counterexample is all it takes to invalidate it. The goal of the designer should be to collect as much pertinent data as possible whereby a false hypothesis is not accepted. This enables the discovery of the truth such as the reason for a product failure. This paradigm is especially true in the world of medicine wherein a larger sample size invalidates the hypothesis that was accepted based upon a smaller sample.
6 Avoid tunnel vision: A designer should always be open to suggestions and new input in order to obtain optimal results. Tunnel vision implies being set in one's ways and not being open to new alternatives. The design process should solicit a range of design alternatives from which the best solution can be picked. Activities such as brainstorming which encourage teamwork and team input should be incorporated.

When a product is designed, it must meet six design requirements. These are listed as follows (Pye, 1989):

1 It is essential that the design accurately embodies the basic principle of arranging the components.
2 The components of any device should have a geometric relationship with one another and to other objects in the product in any way that suits that particular arrangement.

3 The components should have the requisite amount of strength to transmit as well as resist forces as intended.
4 Components should be accessible. This is especially important when the product is designed for ease of disassembly and maintenance.
5 The cost of the final design should be feasible. As far as sustainable product design is concerned, this requirement addresses the "economic" pillar of sustainability.
6 The appearance of the product should be aesthetically pleasing. This is particularly important in the case of consumer products.

1.2.1 The Engineering Design Process

The structure of the overall process of engineering design is quite similar in nature to that of engineering problem solving. In fact, "designing" a product can be likened to a special form of problem solving. According to Hall (1968), the engineering design process consists of the following distinct activities:

1 Defining the problem: This step consists of coming up with a specific definition of the problem that needs to be solved. The different needs and environment under which problem solving needs to be done are analyzed.
2 Design of the value system: Once the problem has been clearly defined, the various objectives (which need to be achieved) as well as criteria need to redefined. This goal is accomplished during this process.
3 Systems synthesis: Just like the engineering problem solving process, the engineering design process is iterative in nature. The optimal design parameters are obtained by studying a range of designs. Several design alternatives need to be considered in order to accomplish this. Multiple design alternatives are generated during the process of systems synthesis.
4 Systems analysis: Systems analysis can be considered as an extension of systems synthesis. While several alternative designs are generated during the systems synthesis stage of the design process, they are analyzed in depth during the stage of systems analysis. Special consideration is given to benefits and disadvantages of each design from the point of view of several design parameters such as functionality, usability, ease of assembly and disassembly and so on.
5 Selection of the best system: Each design problem needs to be solved under a set of specific criteria. These can be likened to constraints. After several designs have been analyzed for particular design objectives, they are next analyzed against the pre-defined criteria during this step.
6 Planning for action: All the analysis that was conducted in the first five stages of the process ultimately leads to a final design choice. The final components of the design are chosen during this step. It consists of the selection of materials for various components, manufacturing processes, dimensions, spatial alignment, dimensions, tolerances and so on.

The next section of the chapter will deal with factors that influence material choice. This is often a function of machining processes, which will be presented in the following section.

1.3 SELECTION OF MATERIALS

Properly selecting a material is crucial to the product design process. It should be obvious to the reader that every product is made from some combination of materials. What material is selected in order to build the final product depends on a variety of factors. In order to arrive at a final decision, several pertinent questions need to be answered. They are listed as follows:

- What is the intended function of the product?
  This question can be answered as follows: The product is intended to insulate. Or, this product is intended to damp vibrations and so on.
- What is the workability of the material?
  Workability of a material implies the ease with which a particular material can be manufactured utilizing conventional machining processes.
- How much is it going to cost to incorporate this particular material in our product?
  Total amount of money spent on acquiring and using the material in a product is a function of direct material cost. (We will address the concept of product costing in the last chapter of this book.)

The issue of workability of materials is dealt with in detail in this section. When deciding upon the appropriate material for a product, workability is intimately connected to the choice of manufacturing process that will be used. The manufacturing process not only imparts the initial form to the material (primary manufacturing processes such as casting, forging, etc.) but also shapes the material in its final form. The latter type of process is referred to as the secondary manufacturing process. Examples of such processes include turning, milling, gear hobbing and so on. Secondary manufacturing processes are also sometimes used to enhance certain material properties such as improving hardness.

To expand upon this basic tenet, there are three main types of manufacturing processes. They are classified as: primary processes, secondary processes and tertiary processes.

- Primary manufacturing processes: As is implied by their name, primary manufacturing processes are used to generate the main shape of the material. Examples of such processes include forging, casting, injection molding and so on. Thus, steel may be cast into ingots using casting. The ingots represent the primary shape of the material.
- Secondary manufacturing processes: Secondary manufacturing processes are used to generate and refine the main shape, form and features of the part. Examples of such processes include machining, grinding and broaching. Turning and milling are further examples of machining processes.
- Tertiary manufacturing processes: After having been subjected to primary and secondary manufacturing, the material is now in its final shape and is very likely

to incorporate most of its intended features. Tertiary manufacturing processes, being unlike their primary and secondary counterparts, have no impact upon the geometry of the part. Such processes are also known as finishing processes and are used for that purpose. Surface treatments and heat treatments are examples of part finishing processes. Only a specific part attribute is affected by most tertiary manufacturing processes.

The functional and utility value of a product is enhanced when certain surface features such as depressions, uniform cross sections and draft free surfaces are incorporated into the product design. A particular type of manufacturing process is required to form such features. Depressions and uniform cross sections, for instance, can be easily formed using sand casting. However, these types of cross sections are not easily compatible with rotational molding. We have presented some guidelines here that are intended to facilitate the material selection process:

1 Materials should be grouped based on their process compatibility:
   - Principal shape-generating processes are used in discrete parts manufacture. Materials should be grouped based on their relationship to these processes.
   - When conducting an initial material search, it is not necessary to search for a particular material. This practice results in slowing down the search. It is also not relevant to making an early decision on the process or material.
   - A super material is defined as a material that possesses the best attainable properties of a material in its class. It is always better to have a super material for each process. Once the super material is obtained, material tradeoffs are then made within the process.

2 The final choice of material depends upon the availability of detailed part specifications. Once these specifications have been obtained, the additional options of materials and processes can be discarded. The ultimate choice is narrowed down to a very limited number of materials and processes. To facilitate this process, computer-aided material and process selection (CAMPS) software may be used.

   In general, material related costs can be minimized by adhering to the following rules:
   - In factory, operations can be minimized by using mill forms that are commercially available. Wherever possible, standard stock shapes, grades and gauges or formulations can be used. It has been observed that larger or heavier selections of a standard material are available for cheaper than smaller or thinner sections of a thinner material. Processing cost and time can be minimized using standardization.
   - Cost of surface finishing operations on a finished component can be minimized by using pre-finished material.
   - Materials should be selected in accordance with two factors: functional requirements and workability. The choice of material should be such that it enhances workability. For instance, free machining grades should

be selected for machined parts. Similarly, easily formable grades should be selected for stamping.

- The part design should maximize material utilization. For instance, designing square ends or features that are nestable with other pieces are examples of such design decisions. Designs that are shown to possess high scrap rates should be avoided.

Design guidelines for cold finished steels are presented here:

- Once the function of a part has been determined, the simplest cross-sectional shape that is consistent with the function should be used. It is recommended that if possible, features such as holes, grooves undercuts and so on should be avoided. Incorporation of such features leads to an increase in manufacturing cost.
- As far as possible, standard shapes should always be preferred over special shapes in order to curb manufacturing cost.
- Because it is difficult to manufacture sharp corners, they should be avoided.
- Unless the design allows for incorporation of generous bottom radii, it is infeasible to use grooves that are deeper than one and a half times width.
- Stress concentration is introduced when a section changes abruptly. For this reason, abrupt changes in section should be avoided.
- Although seamless tubular cross sections are more aesthetic, they are les feasible than welded cross sections. For this reason, seamless cross sections should be avoided.

1.4 SELECTION OF MACHINING PROCESSES

Physical characteristics are imparted to a component part by using specific machining processes. Processing cost and time can be substantially reduced if the correct machining process is selected together with the correct material. Thus, it is crucial to select the correct machining process early.

Standard design guidelines that help in selecting the appropriate machining process are detailed here:

1 Adherence to standard parts and processes
   - As far as possible, standard components should be used. For instance, Toyota uses only about 90% new components from one model to the next.
   - If possible, the work piece should be pre-shaped using a variety of processes such as casting, forging, welding and so on.
   - If possible, standard pre-shaped work pieces should be used.
   - Standard machined features are easier to machine. As such, they result in reduced machining time and cost. Special machined features such as sharp corners are difficult and more expensive to machine. For that reason, they should be avoided.

2 Selection of raw material
   - Raw material should be selected in such a way that it minimizes component cost. This cost includes material cost as well as manufacturing cost.

- Raw material should always be used in the standard form in which it was obtained.

3 Overall design of the component
- The component design should be such that it can be easily machined on one machine tool. Work in process (WIP) inventory is reduced by adopting this practice. In addition, other work elements such as travel times between machines and set-up time are reduced.
- Most machining processes require that the work piece be held in a work-holding device. The component design should be such that the unexposed surfaces of the work piece should not require machining. Only the exposed surfaces should require it.
- Every product design and manufacturing company should play to their strengths. Machine features should be designed in keeping with this attribute. Doing so ensures that the product can actually be manufactured. Otherwise, the company would have to depend on the capability of other firms to accomplish the manufacture of the product.
- The component needs to withstand large amount of forces during the process of machining. In order to be able to do so, the component design should enable it to be held stable in the work-holding device during machining.
- The features should be designed in such a way that when they are to be machined, there will be no interference among the tool, the tool holder, work and work-holding device.
- Any auxiliary holes or main bores that are to be incorporated in the part design should be cylindrical. Also, standard drills and boring tools should be able to easily machine them. This is accomplished by imparting to them appropriate L/D ratios.
- If blind holes are incorporated into a part design, their ends should always be conical. If the blind hole is tapped, it should be ensured that the thread does not continue to the bottom of the hole.

4 Design of rotational components
- Rotational component design is facilitated by ensuring that all plane surfaces are normal to the axis of the component. Similarly, all cylindrical surfaces should be concentric.
- Diameters of internal features should decrease from the exposed face of the work piece.
- As far as internal corners of the component are concerned, radii should be specified that should be equal to the radius of a standard rounded tool corner.
- Long components are often slender and can suffer from fatigue failure. For this reason, incorporation of internal features in long components should be avoided.
- Components should not incorporate very large or very small L/D ratios. The only exception to this rule is if such ratios are functionally important.

5 Design of nonrotational components
- Design of components should be such that a base is provided for the purpose of work holding as well as act as a reference.

- All surfaces of the component that are exposed should consist of a series of plane surfaces that are mutually perpendicular. These surfaces should also be parallel and normal to the base. This is assuming that such part geometry does not interfere with the functionality of the product and other aspects of the part design.
- The internal corners of the product should be designed in such a way that they are normal to the base. The radius of each corner should be equal to that of a standard tool. As far as machined pockets are concerned, the internal corners that are normal to the base should be designed with as large a radius as is possible.
- Plane surface machining should be restricted to only one surface of the component. The different number of component setups and associated setup times are reduced by adhering to this practice.
- Structural rigidity of long components is jeopardized by the incorporation of cylindrical bores. For this reason, such a feature should be avoided in such components.
- Material can be pre-formed to the cross section that is required of a product. Doing so aids in avoiding machined surfaces on long components.
- It is difficult to machine on the inside of components that are shaped like a cube. For this reason, such features should be avoided when designing cubic components.

6 Design of the assembly
- A component that is part of an assembly of components should always be designed in such a way that it is actually possible to assemble it. The concept of design for assembly and disassembly will be dealt with in depth in future chapters of this book.
- It is necessary that every operating machined surface mate with a corresponding machined surface on the mating component.
- Parts should be designed such that internal corners on one component cause no interference with a corresponding external corner on the mating component.

7 Surface finish and accuracy
- Every operating surface is imparted the intended performance by incorporating the widest tolerance and roughest surface finish that is acceptable for such. A significant amount of machining time, cost and effort is saved as a result of this practice.
- Surfaces that require finish grinding should always be raised and they should not intersect to form internal corners.

1.5 WHAT IS DESIGN REVIEW?

Design review (DR) is a systematic process which facilitates bringing products to market quickly. The contemporary environment that is characterized by increasing product complexity makes this process especially relevant. While DR focuses on reducing the time to market, it simultaneously ensures that product design quality is reflective of customer needs. It makes certain that customer needs are met, and the process occurs within the twin constraints of cost and time.

There are two distinct types of design reviews: Formal and informal. Design reviews are incorporated in the product design process so that quality can be built into the product at an early stage. Utilizing DR early in the product design process results in cutting inspection and control costs later. Thus, significant amounts of money, material and energy can be saved.

The formal design review is distinguished from its informal counterpart by the fact that it incorporates standard policies and procedures. Also, each review plays a crucial role in the process of product development and planning. It is not just a perfunctory activity. Specific days are designated just for carrying out DR activities within each development phase. Consistent quality is ensured as a result of using the formal design review.

The informal design review, as its name suggests, is less formal. Unlike its formal counterpart, specific days are not designated just so that this design review can be carried out. The informal review is developed and conducted by individual design reviewers. It is used if and when the need for it arises. It can be used at any stage of the product development process. Resources are not specifically demarcated for this purpose. The effectiveness of the informal design review can vary.

Given the fact that the design stage is the early stage of the product development process, design review is usually applied at this stage. It is applied to a lesser degree at the detailed design as well as the pilot building stages.

Product development for DR consists of four phases as enumerated here:

- Planning stage. This is also referred to as the planning and definition stage.
- Design stage. During this stage, the general and detailed design, prototype, tests and revisions are carried out.
- Production stage. During this stage, the process is designed and pilot production as well as full production is carried out.
- Marketing stage. This stage consists of sales, service and customer feedback.

The following two premises form the foundational blocks for design review:

- Utilize knowledge and technology available from in house and external resources to the greatest possible extent.
- Problems should be tackled and solved as they arise. They should not be passed downstream. In other words, as far as problem solving is concerned, avoid procrastinating.

The steps for the implementation of DR are enumerated as follows:

- Collection and compilation of information.
- Defining quality targets.
- Evaluation of product and process designs as well as supporting operations.
- Proposition of improvements.

- Definition of subsequent actions.
- Confirmation of readiness for the next stage.

Design review does not create good ideas. Rather, it makes good ideas work better. This implies that the primary role of design review is to ensure that good designs are actually manufactured.

The DR process has many advantages. However, its results are sometimes found to be lacking in terms of positive results. This can be attributed to the following reasons:

- It is essential that for a DR to be highly effective, the team responsible for conducting the DR is evenly matched in terms of skills. If this is not the case, results of the DR can be found to be lacking.
- The DR process can produce less than expected results if product developers do not fix mistakes as and how they occur. Mistakes can happen if there is no communication between related departments. Thus, a lack of communication between product development and manufacturing can result in a design that cannot be manufactured. On the other hand, a lack of communication between product development and marketing can result in a product that fails to meet customer requirements. It is easy to visualize many situations of this nature.
- The DR process may suggest changes to the product design. But changes can be made only if there is enough time in the schedule to make such changes. A product development schedule that is cramped will not allow for this to happen. Consequently, the DR process ends up producing subpar results.
- It is essential that members of the DR team possess the requisite quality and quantity of experience. This is critical for the DR process to work effectively.
- DR can have very little impact during the concept phase of the design process. This is because very little feedback can be offered during this stage.
- It is important for design managers to understand and recognize the importance of the DR process. It should be considered as an integral part of the product design process and not separate from it. DR of drawings and specs cannot be done effectively if this fact is ignored.
- It has been observed that the DR process tends to require a substantial amount of work from both design and engineering. Because of this reason, neither of them seems to have much time for the process.

1.5.1 Soft-Hard (S-H) Review

Products need to be designed for safe use in the real world. This objective is achieved by the S-H review. As its name implies, the S-H review comprises two reviews: the soft review and the hard review. The objective of the soft review (S) is to consider the careless misuse of products by users. Careless misuse goes beyond what is represented by normal wear and tear. On the other hand, the objective of the hard review (H) is to consider the loss of function that occurs as a result of malfunction or deterioration over service life of each component. The results obtained from S-H review are applied to product design at an early stage. Future problems with respect to what-if situations of failure and product malfunction can be addressed in time as a result of this practice.

The S-H review has the following advantages:

- It is important to be able to identify quality requirements easily. It is also important to compile resource data. Both objectives are accomplished by studying the different ways in which a product has been used in real life (also referred to as "use in the field").
- Extant general safety regulations and standards may not adequately address every product use situation. This is especially true in the case of soft (S) factors. This is because the "S" factors address unreasonable use of a product. For instance, consider a bicycle rider who jumps off a cliff riding his bicycle. It is doubtful whether the resulting injuries would be sufficiently addressed by the safety regulations that exist on the books. Such gaps are filled by the S-H review.
- Potential product breakdowns and their catastrophic impact can be avoided by incorporating backup systems into product design. These systems are intended to act as a safety measure to mitigate the harmful effects of product failure. A detailed study of component safety functions is essential to the design and fabrication of such systems.

To wrap things up, the S-H review enables the designer to obtain a detailed and clear understanding of applicable safety requirements. It is important that this knowledge be acquired at an early stage of the product design process. It is also possible to utilize the techniques described in the S-H review during the postproduction phase, that is, after the product has been designed. Doing so enables the designer to "continuously improve" the design by comparing actual results with intended ones. Questions such as "How well did the product perform?", "Did it perform to expectations?" and "Did it do better or worse and if so, how?" can be answered.

1.6 DESIGNING A PRODUCT FOR FUNCTIONALITY

Product function is the main objective of product design. Every product is designed to perform a specific function. No matter the level of sophistication contained in the details of the design, it is a fact that designs that are lacking in function exhibit lackluster sales. In other words, products that do not do what is expected of them do not sell. There are many examples of products that are supposed to contain large number of sophisticated features. Yet, somehow, they fail to perform the intended functions. If they do perform the function, they do so unsatisfactorily.

Take the example of the ubiquitous can opener that is found on supermarket shelves. The main function of the can opener is "to open" the can. This function is accomplished by "cutting" the lid of the can. The cutting edge of the can opener moves around the lid in order to sever it entirely and cleanly. This process should be accomplished without leaving slivers of metal behind. That is the ideal objective.

But in reality, this is rarely the case. Most can openers routinely fail to perform the intended function. Not only are they not able to cut the lid in order to open the can, the cutting action "jiggles" the lid. The contents of the can are splattered as a result of the "jiggling." As the cutter progresses around the can, it can also submerge the lid in the contents of the can.

It is important to know what is the intended functionality of the product being designed. This knowledge is required in a whole host of design-related activities such as:

- Generation of designs and their modification.
- Comparison, evaluation and selection of designs and their diagnosis.
- Product repair.

Most designers and researchers tend to agree that function is an important concept and plays a critical role in determining a product's fundamental characteristics. However, a clearly defined objective and uniformly applicable and widely accepted definition of functionality is largely absent. Historically, function has been interpreted in numerous ways. Examples of this interpretation include the following.

- The intended behavior of the design is abstracted.
- The intended behavior of the design is indexed.
- Function is expressed as a relationship between the environment and the design.
- Function is expressed as an external behavior of the design.
- Function is expressed as the internal behavior of a design (Umeda and Tomiyama, 1997).

Designing a product for functionality follows a systematic six-step process. Each step of the process is described in brief as follows:

1 Extending the definition of functionality:

In this step, the definition of functionality is extended. As a starting point, we define the word "function" as the ability of a product to do something reliably, safely and in a usable manner such that it can also be easily manufactured with a concern for the environment.

2 Development of specific functionality criteria:

A generic terminology in the form of a checklist is generated for particular product design and manufacturing guidelines. This is important since different products have different functions and manufacturing processes. This renders the development of a list of generic terminologies across a variety of consumer products difficult if not impossible. The criteria of performance, manufacturability, safety, environment friendliness and reliability were specifically considered during this step. Important factors pertaining to specific functionality criteria are identified in each of the afore-mentioned attributes that need to be controlled through design variables, material variables and manufacturing process variables. Doing so would render the product functionally superior.

3 Classification of each criterion into different design stages:

This step helps remedy one of the principal drawbacks present in most DfX tools. These tools are not always very clear as to when they should be used or how they should be used. Most of the time, they suggest a list of design rules with little to no direction as to their practical use. This step utilizes the broad phases of task clarification, conceptual design, embodiment of design and detailed design. The methodology presented in this chapter uses the overall guideline development procedure.

4 Systematically identifying design and manufacturing variables that affect functionality:

Design is inherently a human activity. People design things. Design variables are based on specific functional requirements. The objective is to find out whether the requirements are satisfied by the design variables. Design variables are affected by a variety of factors such as experience of the designer, types of design tools used, type of design (such as creative design or adaptive redesign), design budget and the type of communication mechanism used (such as concurrent engineering vs. conventional over-the-wall approach).

Manufacturing variables comprise both materials as well as manufacturing processes. We have already discussed each of these in some detail at the beginning of this chapter. The material selected by the designer should satisfy the functional requirements of the component. This is the main concern insofar as material selection is concerned. The designer should be knowledgeable in terms of what properties need to be considered, how they are to be determined as well as limitations that should be placed on the application. Variables such as the type of material, material toughness, fatigue resistance and hardness resistance significantly affect product function. We have also mentioned that the choice of material, in turn, plays an important role in choosing the appropriate machining process as well. Manufacturing process dimensions such as formability, weldability, machinability and ease of assembly are thus determined by the choice of material. Product and component functionality are optimized by controlling process variables. Examples of such variables include duration of machining, current density, voltage, type of solvent used and its amount, cutting speed, cutting feed, depth of cut and so on. Guidelines presented here are based on manufacturing variables only.

5 Determining the relationship between product functionality and design and manufacturing variables:

In this step, guidelines are generated with the objective of ensuring functionality of a product. Design guidelines help determine what variables need to be controlled. This is the case if the generic design criteria are obvious. On the other hand, if they are not obvious, the relationship is demonstrated with the help of a different set of variables. Controlling the design and manufacturing variables optimally helps ensure functionality of the product. Overall product functionality is enhanced by developing a link between product functionality and design attributes on the one hand and product design and its manufacturing attributes on the other hand. We will utilize the four-house quality function deployment (QFD) approach to help demonstrate this methodology.

6. Systematically developing design and manufacturing guidelines:

This step involves developing design and manufacturing guidelines. These guidelines use the links outlined in step 5 above. The main function of the guidelines is to control the design and manufacturing variables that control product functionality. The guidelines also serve to control any major design activities that pay an important role in the design process.

It is important to note that the aforementioned six steps are iterative in nature. They are also open to receiving new information. Any new relevant concepts and information can be added to the process as and how they become available. The main hypothesis being tested is that the new guidelines and criteria that have been designed for the purpose of ensuring functionality do indeed achieve that result.

In the next section of this chapter, we present two case studies. The first case study is that of a can opener and it deals with product functionality. The second case study is that of a toaster and it deals with product usability. Each case study has two parts. The first part presents the four-house QFD approach while the second part presents the development of design checklists to achieve the intended design goal (functionality for a can opener and usability for an electric toaster).

1.6.1 Case Study: Can Opener

1.6.1.1 Guideline Development for Can Opener

Functional design guidelines can be developed for any product using a generic checklist approach. An example of such an approach is presented in this section. Consider the case of a can opener. What are the functional requirements for a can opener? How can these be converted into generic design guidelines?

Specific generic criteria are applied to develop guidelines for the can opener. The following specific criteria are considered, and the corresponding functionality is presented in parentheses. Thus, while redundancy is a design characteristic, reliability can be construed as a functional requirement. Similarly, user friendliness is a design characteristics and usability is the corresponding functional requirement. This means that if user friendly design features are incorporated into the can opener, it will render the can opener usable. If the correct material and processes are selected, the can opener can be easily manufactured.

- Functional effectiveness (performance).
- Environment in which the product operates (performance).
- Redundancy of design (reliability).
- User friendliness (usability).
- Avoiding sharp corners (safety).
- Attractiveness (quality).

- Selection of materials and processes (manufacturability).
- Avoidance of toxic materials and ease of recycling (environment friendliness).

1.6.2 What Are the Manufacturing Attributes That Affect Functionality?

Functionality of the can opener is based upon functional analysis and user requirements. The following can be considered to be the principal functional requirements of a typical can opener:

- Lid should be cut smoothly.
- Operation can be accomplished smoothly.
- The process of cutting should be easy.
- The device can be handled safely.
- The can opener should offer a comfortable grip.
- The device should be durable and possess an attractive appearance.
- The device should meet kitchen needs.
- The device should be hygienic.

The following is a discussion of critical functional features and the corresponding manufacturing processes.

- Cutting should be easy: The ease with which the lid can be cut is a function of the hardness of the blade. Appropriate material selection and heat treatment directly influence the hardness of the material. The quenching method and temperature are the primary process variables that control the effectiveness of heat treatment. Other parameters that are associated with heat treatment can be easily controlled as well. Examples of such parameters include decarburization, scaling, cracking, residual stresses and dimensional changes. These parameters can be selected and controlled in order to optimize the end results of the process.
- Lid should be smoothly cut: Smoothness of the cut edge is associated with the sharpness of the cutting blade. A sharp cutting edge can easily rip through the metal. Swaging of the blade's edge can impart a high degree of sharpness to the blade. Tensile strength and surface hardness are improved by the swaging process. This is due to the better flow of metal and a finer grain structure.
- Operation should be slip free: Slip free operation is achieved by a tight and smooth roll of the drive sprocket on the outside of the rim of the can. How easily the sprocket roll is able to move is influenced by the hardness and wear resistance of the drive sprocket. Heat treatment and nickel plating can vastly improve these properties.
- Durability and appearance: Surface appearance and durability are enhanced by surface preparation and coating. Appearance is improved by surface preparation while coating improves durability.

1.6.3 Development of Linkages to Connect Functionality with Manufacturing

The four-house QFD process is presented in this section to establish a relationship between functionality and manufacturing. Functional requirements are identified and translated into design and technical requirements during the conceptual design stage. Product deployment constitutes the second stage of the transformation process. The previously developed design and technical requirements are transformed into product specifications and features during this stage. Various product features are converted into specific manufacturing operations during the process deployment stage. Different manufacturing processes are related to process and material variables (which control them) during the manufacturing deployment stage.

1.6.3.1 QFD House 1: Transforming Functional Requirements into Technical and Design Requirements

The first house of quality transforms the functional requirements of the product into design and technical requirements. The functional requirements are based on the extended definition of function that has been presented in the previous pages. The first house of quality (HoQ) is depicted in Table 1.1.

1.6.3.2 QFD House 2: Product Deployment—Converting Technical Requirements into Product Features

During this step, technical characteristics identified during the first design stage are transformed into product features. The product features are designed in such a way that they most efficiently satisfy the technical requirements. Each individual product feature is clearly identified in vertical columns. The second HoQ is presented in the Tables 1.2, 1.3 and 1.4.

TABLE 1.1
Transforming functional requirements into technical design features (HoQ1)

Relationships **Strong (5): @** **Medium(3): #** **Weak (1): $**	**Design and technical requirements**	**Structural rigidity**	**High cutting force of blade on lid**	**High contact force of sprocket on rim**	**Low handle grip force**	**Low crank torque**	**Smooth surface and edge finish**	**Good rust proofing**
	Performance	@	@	@	#	#		
	Reliability	@	#	#	$	$	#	#
	Manufacturability	#	@	@			#	#
Functional requirements	Safety	#					#	$
	Usability	@	@	@	@	@	#	$
	Quality	#					@	@
	Environment friendliness		$				$	$

TABLE 1.2
Transforming technical requirements into product features (HoQ2)

Relationships **Strong (5): @** **Medium(3): #** **Weak (1): $**	**Product features**	**Blade hardness**	**Drive sprocket hardness**	**Material**	**Surface and edge regularity**	**Crank surface and edge regularity**	**Surface finish**	**Joints between components**
Design and technical requirements	Structural rigidity		@					@
	High cutting force	@	@	@				#
	High contact force		@	@				#
	Low grip force	@	@	@	$	$		@
	Low crank torque	@	@	#	$	$		
	Smooth surface and edge finish			#	@	@	@	
	Good rust proofing	#		#	#	#	@	

1.6.4 Functionality Evaluation Test: Development of a Survey

The functionality evaluation and design/manufacture guidelines presented in the previous section were tested using two surveys. The surveys were in the form of questionnaires. Specifically, two sets of questionnaires were developed. The first questionnaire evaluated user perception and the second questionnaire evaluated attributes related to designer/manufacturing engineering. Twenty-one subjects participated in the user evaluation. Their responses were obtained through a one-on-one interview. All the subjects had previous experience using can openers. They owned manual can openers and were thoroughly familiar with what the device was supposed to accomplish. The subjects rated each answer on a scale of 1–5 (from least important to the most important).

As for the second objective, guideline questionnaires pertaining to design and manufacturing were either faxed or mailed to respective designers and manufacturers. Designer profile forms were used to collect background information on each designer. Analysis was conducted on the completed questionnaires and the results were studied. Examples are presented in Tables 1.5, 1.6 and 1.7. The tables present checklists pertaining to performance, reliability, safety and other aspects of functionality. It is found that in order to be fully functional, a product needs to "perform" the intended function "reliably" and "safely." Since we are dealing with sustainable product development, the checklist corresponding to "environment friendliness" is also presented.

Design and manufacturing guidelines for performance and environment friendliness are presented in Tables 1.8 and 1.9.

TABLE 1.3
Transforming product features into process features (HoQ3)

Relationships Strong (5): @ Medium(3): # Weak (1): $	Process features	Blanking (upper handle)	Stamping (upper handle)	Blanking (lower handle)	Stamping (lower handle)	Blanking (blade)	Swaging (blade)	Heat treatment (blade)	Blanking (crank)	Stamping (crank)	Stamping (drive sprocket)	Heat treatment (sprocket)	Tumbling (edge and surface)	Nickel plating	Swaging (subassembly)	Swaging (subassembly)	Riveting (overall assembly)
	High blade hardness					$	#	@									
	High drive Sprocket hardness										$	@					
	Smooth handle surface and edge	@	#	@	#								@	#			
Product features	Smooth crank surface and edge								@	#			@	#			
	Good surface finish												$	@			
	Strong joint between parts														@	@	@

TABLE 1.4
Transforming process features into process variables (HoQ4)

Relationships Strong (5): @ Medium(3): # Weak (1): $	Process Variables	Pressure	Temperature	Chemical concentration	Duration	Quench solution	Operation speed	Punch-die clearance	Rate of cooling	Rivet-hole clearance	Current density
Process features	Blanking	@					#	@			
	Stamping	#					@	@			
	Swaging	@					@				
	Heat treatment		@	@	@	@				@	
	Tumbling			#	#		#				
	Nickel plating			@	#		$				@
	Riveting	@					$			@	

TABLE 1.5
Checklists for evaluation of "performance" and "reliability" of can opener

		Opinion					
No.	**Question: Performance**	**1**	**2**	**3**	**4**	**5**	**Comments**
1	To what extent are you concerned with the force you need to apply to open the can?						
2	To what extent are you concerned with the structural rigidity of the can opener when you open the can?						
3	To what extent are you concerned with the force needed on the lower/upper handle to make blade pierce the lid?						
4	To what extent are you concerned with comfortable feeling of the grip of the handlebars during opening of the can?						
5	To what extent are you concerned with the lid or the can having sharp or jagged edges?						
6	What is the overall functionality rating of the can opener in terms of its design for performance in this section?						
No.	**Question: Reliability**						
1	To what extent are you concerned with the blade failing often to pierce the lid?						
2	To what extent are you concerned with the guiding wheel failing often during opening the can?						
3	To what extent are you concerned with the joint (rivet) failing (breaking) often?						
4	To what extent are you concerned with the joint loosening (loss in structural rigidity) often?						
5	To what extent are you concerned with the nickel plating of the can opener wearing out too soon (corrosion resistance and comfortable grip)?						
6	What is the overall functionality rating of the can opener in terms of its design for reliability in this section?						

TABLE 1.6
Checklist for evaluation of "safe performance" of can opener

No.	Question: Safety	Opinion 1	2	3	4	5	Comments
1	To what extent are you concerned that the burrs on the upper and lower handle can pose a safety problem to the user?						
2	To what extent are you concerned that the can opener leaves a sharp and jagged edge on either the can or the lid?						
3	To what extent are you concerned that the can opener should provide guards or covers oversharp blades and similar elements?						
4	To what extent are you concerned that the can opener should be free from sharp edges, corners and points that can cause injury?						
5	To what extent are you concerned that the can opener should prevent toxic contamination of metal shaving from falling into the food?						
6	To what extent are you concerned that the cutter wheel or blade never comes in contact with the food?						
7	What is the overall functionality rating of the can opener in terms of its design for safety?						

TABLE 1.7
Checklist for evaluation of "environment friendliness" of can opener

No.	Question: Environment friendliness	Opinion 1	2	3	4	5	Comments
1	To what extent are you concerned that the can opener should avoid as much as possible the use of toxic materials in the product and manufacturing process?						
2	To what extent are you concerned that the can opener should avoid the use of hazardous materials including those that are a hazard when burned, recycled or discarded?						
3	Do you agree that the can opener should minimize the amount of material in the product?						
4	What is the overall functionality rating of the can opener in terms of its design for environment friendliness						

The validity and reliability of the checklists is determined using standard statistical testing techniques. In the next section of this chapter, we discuss the concept of product usability in detail.

1.7 DESIGNING A PRODUCT FOR USABILITY

Usability implies the ease with which something can be used. In our case, a product that is designed for usability is likely to be easy to use. Consider the case of two

TABLE 1.8
Design and manufacturing checklist for "performance" of can opener

No.	Design and manufacturing guidelines for performance	Opinion 1	2	3	4	5	Comments
1	Designer of the can opener should select appropriate material that meets the functional requirements of the users along with various product features						
2	Design should consider low crank torque and low handle grip force to open the can						
3	For structural rigidity and durability, use two rivets to join the upper handle and lower handle						
4	Design should consider force analysis. The cutting edge of the blade and the crank of the guiding wheel should have a shear stress greater than the maximum material strength of the lid						
5	Design should consider all components strong enough to transmit and resist forces during opening of the can						
6	Removing the burrs results from stamping operations on the upper handle, lower handle, blade and crank is important. Burrs can cause jamming of parts, reduce the fatigue life of components and can be a safety hazard						
7	Design of the blade should consider heat treatment so that it never leaves a sharp or jagged edge on either the lid or the can						
8	What is the overall functionality rating of the can opener in terms of its design for performance?						

TABLE 1.9
Design and manufacturing checklist for "environment friendliness" of a can opener

No.	Design and manufacturing guidelines for environment friendliness	Opinion 1	2	3	4	5	Comments
1	The use of toxic materials should be avoided to the greatest extent possible in the product as well as the manufacturing processes						
2	Use of hazardous materials should be avoided in product design						
3	The use of material should be minimized in the product design						
4	Process should be designed so as to minimize manufacturing residue such as mold scrap and cutting scrap						
5	Design the fasteners of the can opener for easy access to aid in disassembly						
6	Minimize using liquids such as acids, alkalis and solvents during the manufacturing process						
7	What is the overall functionality rating of the can opener in terms of its design for environment friendliness?						

almost identical products. They are identical from the point of view of functionality. However, they are not equally usable. While one of the products is very user friendly, the other is quite cumbersome to use. The consumer is likely to prefer and even pay more for the one that is user friendly.

There are many examples of design features that enhance product usability. These include ergonomically designed operation knobs, controls that can be operated using minimal force (such as soft touch buttons), swiveling chairs, spanners with swiveling pivots that enable easy access into corners and so on.

A product designed for usability serves a dual purpose. On the one hand, it pleases the user while simultaneously bringing in more revenue through increased sales. Figure 1.1 relates the concept of usability with that of design and manufacturing.

User friendliness of a product is enhanced if the functions performed by humans are within their capability. In addition, it is important that the product user interface is physically comfortable and does not cause mental stress (Haubner, 1990; Nielsen, 1993). A user friendly system should be easy to learn, easy to recall and comparatively error free (Nielsen, 1992, 1993).

Reliability is an important aspect of product design. It enhances the degree of usability of that product. Reliability of a product is the likelihood that under normal working conditions, a product will perform satisfactorily for a particular period of time (Anderson, 1991). Reliability is measured in terms of mean time to failure (MTTF). MTTF is defined as the average of mean lifetime for a particular population of products.

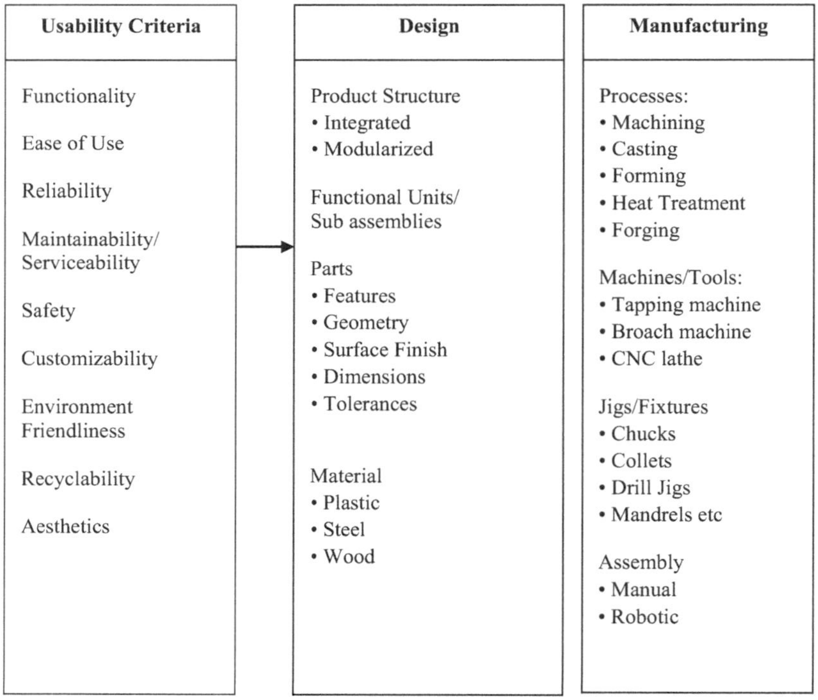

FIGURE 1.1 Design criteria and manufacturing factors to design products for usability.

Continuous improvement of materials, product design, manufacturing processes and use environment leads to improved product reliability (Alonso, 1990; Comizolli et al., 1990). Reliability growth test management is a critical component of the product assurance function (Bieda, 1992). The reliability growth during the product development process is quantified and monitored using computer applications such as knowledge decision support systems (Nasser and Souder, 1989).

To better understand the concept of product usability, let us consider the case of a toaster. We will use the twin approaches of QFD and design checklists to help readers grasp the basic tenet.

A toaster applies radiant heat directly to bread slice. When a surface temperature of 310° F is attained, a chemical change known as the Maillard reaction begins (Consumer Reports, 1990). During this reaction, sugars and starches begin to caramelize and turn brown and begin to take on intense flavors that are characteristic of a toast. Excessive heating turns the underlying fibers into carbon, thus burning the toast. The primary function of a toaster is to control the location and amount of heat delivered to a slice of bread.

As before, we will once again use the four houses of quality to gain a better understanding of product usability. This approach will be similar to the one we used for functionality.

The first HoQ is depicted in Table 1.10. This house converts the customer requirements into technical characteristics. The technical requirements are listed in the vertical portion of the QFD.

The customer requirements were translated into appropriate technical requirements as follows.

- The temperature attained inside the toaster should be just adequate to brown the slice. The temperature should not burn the bread. Also, it should not be too cold to turn the color.
- The application of heat should be uniform enough to brown the slice uniformly.
- The width of the toaster slot should be large enough to accept a variety of slice thickness such as sliced bread as well as thicker bagels.
- The time of exposure for the slice to heat should be adequate enough to produce a fine toast.
- It should be possible to press down the lever for a range of customers starting with small children going all the way up to the elderly.
- Controls and different knobs should be located to facilitate easy access.
- The color and gloss of external finish should be pleasant in appearance. The product shape should be appealing as well.
- The toaster should be made of a stain-resistant outer case. It should be easy to clean the casing with water.
- The outer casing and inner frames should have high-impact resistance and resist breakage on falling.
- The toaster should be adequately insulated with the help of thermal insulation.
- The thermal insulation should avoid dissipation of heat. This will lead to conservation of electrical energy.

- The electrical insulation should be good enough and the electrical circuit should never come in contact with the metal frame (which causes electrocution).
- The toaster material should provide adequate structural integrity to the product. It should be easy to reuse or recycle the toaster upon the end of its useful life.

Customer requirements are ranked in order of importance. The relationship between customer requirements and technical requirements are represented in the central matrix of the HoQ. The relationship can be strong, medium or weak (as shown in Table 1.10).

The second HoQ is depicted in Table 1.11. This house transforms technical characteristics into product features. The case study is about a mechanical toaster. It consists of an inner U frame. A steel rod is attached to the other end of the frame. The pushdown lever moves on this rod. A pull force is exerted by a steel coil attached to the toast lever. It always keeps the lever on top. When the lever is pushed down, it lowers the bread holder located inside the frame. A copper rod runs through the

TABLE 1.10
Transforming customer requirements into technical features for electric toaster: Usability study

Relationships **Strong (5): @** **Medium(3): #** **Weak (1): $**	**Technical requirements**	**Importance**	**Uniformity of heat application**	**Duration of heat application**	**Force on toast lever**	**Color of exterior**	**Thermal/electrical insulation**	**Accessibility**	**Material properties**	**Competitive analysis** 1	2	3	4	5
	Functional requirement	8	@	@			$		$		•			
	Ease of use	7			@								•	
	Aesthetics	4				@			#		•			
Customer requirements	Maintainability	3							@			•		
	Reliability	5	#	#	#				#				•	
	Energy conservation	2					@					•		
	Safety	6					@						•	
	Recyclability	1							@			•		
Competitive analysis		1												
		2			•	•								
		3	•				•	•						
		4		•					•					
		5												
Weighting			9	9	8	4	8	3	7					

TABLE 1.11
Transforming technical requirements into product features for electric toaster: Usability study

Relationships **Strong (5): @** **Medium(3): #** **Weak (1): $**	**Product features**	**Importance**	**Thermostat**	**Bread holder**	**Toast lever assembly**	**Color control knob**	**Side and end covers**	**Crumb tray**	**Material of parts**	**Comments**
Technical requirements	Duration of heat application	9	@		$	#				Should be able to caramelize the bread
	Uniformity of heat application	9	#	@						Should be able to produce uniform color on the toast
	Force on toast lever	8			@	$				Low enough to be operated by children
	Exterior finish	4					@		$	Improves appearance
	Thermal and electrical insulation	8					@		#	Exterior should not be hot
	Accessibility	3					#	@		Should be easy to remove crumbs and for cleaning
	Structural rigidity	7							@	Should withstand heat
Competitive analysis		1								
		2	.			.				
		3		.	.		.		.	
		4						.		
		5								
Weighting			11	8	8	4	12	3	9	

bottom of the toaster. One of its ends is connected to a contact switch. This contact switch assembly comprises a heating coil and completes the electrical circuit. The other end of the copper rod is activated by the toast lever as the latter is lowered to the bottom. The heating filament is wrapped on two sheets of mica. These sheets are constrained vertically by attachments located on the U frame.

When the toast lever is lowered, the metal strip that moves against the vertical rod locks it in place. Lowering of the lever also causes the circuit to close. It simultaneously heats the coil. A bimetallic strip is attached to the bottom of the U frame. This strip acts as a thermostat. It expands upon being heated and pushes a rod. This action releases a weight to fall. The falling weight in turn releases the metal strip holding the lever down. On being released, the lever moves up, carrying along with it the weight and causes the bread holder to pop up. The length of the bimetallic strip is controlled by the color control knob. The control knob controls the duration of the heating.

Two side covers and end covers enclose the U frame housing. A crumb tray is attached to the bottom of the frame. Its function is to retain the falling bread crumbs.

The third HoQ is depicted in Table 1.12. This house transforms product features into process features. The end bracket is made of galvanized steel 40 milli-inches thick to impart structural rigidity. A punch press is used to stamp the end brackets and base plate. The brackets are then mounted vertically on both ends of the base plate and are spot welded in place. The assembly containing the coil and the weight is tabbed onto the left end bracket. The vertical rod is clamped between the coil assembly and the base plate. The toast lever is held by the vertical rod. The bread holder rod made of steel is coated with copper. This is done to prevent rusting and contamination with food. Three mica sheets with heating coil wound on them are mounted onto the base plate. The first sheet is the center and the other two sheets flank it. The heating element made of nichrome is wound on mica sheets. Each mica sheet is 15 milli-inches thick. The electrical switch assembly is tabbed to the right side of the end bracket. It is connected to the electric cord. The thermostat assembly is tabbed with the base plate. The crumb tray is made of galvanized steel sheet 18 milli-inches thick. The tray is pivoted with the base plate using a thin steel rod.

The U frame is made of galvanized steel. It is 25 milli-inches thick and is chrome plated on top of a nickel coating. The chrome and nickel coating impart corrosion-proof qualities to the cover. Heat is retained inside the toaster due to high reflectivity of the polish. The U frame and side covers are cold rolled. The end covers are made of thermoset polyester and are plastic molded. Screws hold the entire assembly in place. The steel covers are painted on the outside to enhance appearance. The color of the end covers is chosen so as to meet customer preferences.

The fourth HoQ is depicted in Table 1.13. During this stage, the various manufacturing techniques necessary to manufacture the product are related to process attributes that affect them either positively or negatively. Some of the important processes that are relevant to toaster manufacture are progressive die operations, plastic molding, assembly processes and painting and inorganic coating. Each manufacturing process is listed on the left side of the house. The processes are listed along the rows whereas process variables are listed in the columns.

Questionnaires were developed to test the usability of a product design. These were similar to the ones developed for product functionality. Once again, two sets of questionnaires were developed. The first questionnaire evaluated user perception and the second questionnaire evaluated attributes related to designer/manufacturing engineering. Twenty-one subjects participated in the user evaluation. Their responses were obtained through a one-on-one interview. All subjects had previous experience using can openers. They owned manual can openers and were thoroughly familiar with what the device was supposed to accomplish. The subjects rated each answer on a scale of 1–5 (from the least important to the most important). Generic checklists for attributes such as usability, environment friendliness, aesthetics and so on are presented in Tables 1.14, 1.15, 1.16, 1.17 and 1.18.

TABLE 1.12
Transforming product features into process features for electric toaster: Usability study

Relationships			Progressive die operation				Assembly operation			Painting and inorganic coating				
Strong (5): @ Medium(3): # Weak (1): $	Process Features	Importance	Blanking	Stamping	Tabbing	Compression molding	Spot welding	Screw fastening	Riveting	Painting	Chromium coating	Nickel coating	Shredding/melting	Comments
	Thermostat assembly	11	#	#	#									Assembly purchased from vendor and checked for quality
Product features	Bread holder	8	$	$	$		$				#			Should be rust proof and not contaminate with toast
	Toast lever assembly	8	$	$	$									Pressing toast lever should not tip the toaster off
	Color control knob	4	$		$		$							Color of toast should be consistent for a batch of 6
	Side and end covers	12	$		$	@		$		#	#		#	Covers should be easy to assemble and disassemble
	Crumb tray	3	$	$		$		#	$			$		Should be adequately accessible for easy removal
	Material of parts	9											#	Should have necessary thermal and electrical properties
Cost level		L	.	.	.			.	.	.				L= Low
		M					.				.	.	.	M= Medium
		H				.								H = High
Weighting			7	6	1	11	2	2	2	4	6	2	6	

TABLE 1.13
Transforming processes into process variables for an electric toaster: Usability study

Relationships Strong (5): @ Medium(3): # Weak (1): $	**Process Variables**	**Importance**	**Pressure**	**Temperature**	**Chemical concentration**	**Duration**	**Pigment**	**Speed**	**Punch-die clearance**	**Rate of cooling**	**Rivet-hole clearance**	**Current density**	**Comments**
	Blanking	7	@					#	#				Progressive die operation
	Stamping	6	@					#	#				Progressive die operation
	Spot welding	2										@	Joins frame components
	Tabbing	2	@						#				Joining components
	Screw fastening	2											Manual operation
Process features	Riveting	2	@					$			@		Assembly operation
	Compression molding	11	@	@		#							Purchased from vendor
	Painting	4			#	#	@	#		#			Organic coating
	Chromium coating	6			#	#		#				#	Improves appearance and corrosion resistance
	Nickel coating	2			#	#		#				#	Improves rust proofing
	Shredding/melting	6		@									Material recovery process
		Range											
		Units	KPa	Degree C	PPM	Min/Sec	Kg/Cu.ft	Inch/min	Micro inch	Degree/min	Micro inch	Ampere	

TABLE 1.14
Generic checklist developed for evaluating "ease of use" of a toaster

No.	Usability evaluation for ease of use	Scale: bad 1	2	3	4	5	6	good 7	Comments
1	How comfortably are the controls and displays located?								
2	How obvious are the operations of the controls?								
3	How easy is it for a novice user to intuitively understand the displays and controls of the product?								
4	How well have the control tasks been simplified?								
5	While making the product controls and displays very simple, does the product still have enough complexity to motivate the user?								
6	How well is the possibility of moving the controls in the wrong direction been eliminated?								
7	How well are knobs and handles differentiated?								
8	How adequately is the feedback provided during product use?								
9	How well is information displayed?								
10	How well does the product design anticipate human errors in operation?								
11	What is the overall usability rating of the product in terms of ease of use?								

TABLE 1.15
Generic checklist developed for evaluating "aesthetics" of a toaster

No.	Usability evaluation for aesthetics	Scale: bad 1	2	3	4	5	6	good 7	Comments
1	How good is the color or finish of the product exterior?								
2	How good is the texture of the exterior and interior of the product?								
3	How soft and flexible is the contact surface of the product?								
4	How inviting or attractive is the product shape to the user?								
5	How good is the product in imparting a feeling of cleanliness?								
6	How good is the product design in incorporating intricate and appealing shapes?								
7	What is the overall usability of the product in terms of aesthetics?								

TABLE 1.16
Generic checklist developed for evaluating "reliability" of a toaster

		Scale							
		bad						good	Comments
No.	**Usability evaluation for reliability**	**1**	**2**	**3**	**4**	**5**	**6**	**7**	
1	How well has the design been simplified?								
2	How well are proven, standard or existing components and design approaches being used?								
3	How well is the product protected against environmental factors such as heat, moisture and so on?								
4	How good is the design of the product in failing safe?								
5	How good is the design in permitting continued operation when a critical component fails?								
6	How high is the reliability of the parts or components of the product?								
7	How good is the design in protecting against corrosion and thermal extremes?								
8	How well are the sensitive and weak components protected from damage during use or maintenance?								
9	What is the overall usability rating of the product in terms of reliability?								

TABLE 1.17
Generic checklist developed for evaluating "environment friendliness" of a toaster

		Scale							
		bad						good	
No.	**Usability evaluation for environment friendliness**	**1**	**2**	**3**	**4**	**5**	**6**	**7**	**Comments**
1	How good is the design in removing all toxic materials and incorporating those that are most environmentally preferable for the desired function?								
2	How well does the design avoid the use of materials that are restricted in supply?								
3	How well is the product designed to utilize recycled materials wherever possible?								
4	How well does the product incorporate measures to eliminate the use of or release of green house gases?								
5	How does the product design avoid producing liquid or solid residues whose recycling will be difficult or energy intensive?								
6	How well does the design minimize the use of energy-intensive process steps such as high heating differentials, heavy motors, extensive cooling and so on?								

TABLE 1.17
(Cont.)

No.	Usability evaluation for environment friendliness	Scale bad 1	2	3	4	5	6	good 7	Comments
7	How well is the product designed to minimize the use of materials whose extraction is energy intensive?								
8	How good is the design in minimizing the number and volume of different packaging materials?								
9	How well does the product design avoid using materials whose transport to the facility will require significant energy use?								
10	What is the overall usability rating of the product in terms of environmental friendliness?								

TABLE 1.18
Generic checklist developed for evaluating "recyclability and disposability" of a toaster

No.	Usability evaluation for recyclability and disposability	Scale bad 1	2	3	4	5	6	good 7	Comments
1	How well does the design minimize the number of different materials that are used in manufacturing?								
2	How well does this product minimize the use of toxic materials?								
3	How easy is it to identify and separate toxic materials if used?								
4	How good is the design in avoiding joining dissimilar materials together in ways that are difficult to reverse?								
5	How well does the design incorporate thermoplastics in use of thermosetting materials?								
6	How good is the design in avoiding painting or the use of plated metals?								
7	How good is the design in using fasteners such as clips or hook and loop in assembly in place of bonds and welds?								
8	How good is the identification, by ISO markings of various plastics or other materials as to their content?								
9	How good is the design in minimizing or eliminating fillers?								
10	How good is the design in avoiding assemblies that are difficult to separate?								
11	How good is the design in eliminating materials that are difficult to recycle?								
12	What is the overall usability of the product in terms of recyclability and disposability?								

The validity and reliability of the checklists is determined using standard statistical testing techniques.

We have presented a succinct overview of the product design process in the preceding pages. Do keep in mind that this being a text on sustainable product design, a more comprehensive overview of the general product design process is beyond its purview. The next section of this book will present some basic rules as for sustainable product design.

1.8 GOLDEN RULES OF SUSTAINABLE PRODUCT DESIGN

QFD, design guidelines and design checklists are some of the tools more commonly used in the product design process. We have already encountered several examples of these tools in the previous part of this chapter. Specifically, design checklists and QFD were used when we presented the concepts of product functionality and product usability. Also, design guidelines were used in conjunction with consideration of materials and processes. Information presented in this chapter so far was intended to act as a foundation. Its purpose was to help readers get a thorough understanding of the product design process itself. Future chapters will build upon this foundation. The ultimate goal is to understand the process of sustainable product design in particular.

We are certain that readers have appreciated by now the fact that sustainable product design is a specialized version of the overall product design process. The variation occurs in the types of materials and manufacturing processes that can be used in the creation of a sustainable product. Sustainable products also use different types of fastening methods than their nonsustainable counterparts. Such fasteners and fastening methods enable ease of disassembly and maintenance, which are crucial to the sustainability of a product.

Therefore, this part of the chapter is intended to act as a transition towards achieving sustainable design strategies. With that purpose in mind, we have presented ten golden rules (Luttropp and Lagerstedt, 2006) in this section that can be followed when designing such products.

Rule 1: Avoid the use of toxic substances in product design where possible. Closed loops should be utilized for toxic substances that are unavoidable. The following suggestions are intended to help implement this rule in practice:

- The toxic substances currently being used in the product should be identified.
- It should be investigated whether the toxic substance can be replaced by a nontoxic substitute. The replacement should fulfill functional and economic requirements of the product.
- Materials used in toxic substances are harmful to the environment if they are disposed of carelessly. This can be avoided by utilizing closed loop systems which recycle the material and prevent it from polluting the environment. Whether such loops have already been established should be investigated. If not, the designer should try to develop them.

Rule 2: The amount of energy and resources expended during the production and transportation phase of product development should be minimized. This can be achieved through improved housekeeping.

Consider the example of a part that does not use standard shapes. This leads to several problems such as suboptimal use of raw material, problems with packaging and transportation, not to mention problems with material handling and assembly. Using a standard shape is likely to solve each of the aforementioned problems. However, the changeover to a standard shape should not have any detrimental impact on product functionality.

Rule 3: High-quality materials and structural features should be used with the intent to minimize product weight. This should, however, not interfere with functional requirements of the product such as flexibility, impact strength and so on.

Rule 4: During the usage phase of a product, energy and resource consumption should be minimized.

Energy consumption can be minimized by replacing energy-intensive components with other components that do not consume as much energy. For instance, an engine could be replaced by a winding mechanism.

Rule 5: Wherever possible, practices such as product repair and upgrade should be encouraged. This is especially true for system-dependent products such as computers, cell phones and CD players.

Rule 6: Sustainable products should have a longer life than their un-sustainable counterparts. Products that are designed to last longer are also easier to reuse. For this reason, they do not need to be recycled or even remanufactured. This is especially true for products that have significant environmental aspects outside their usage phase.

Product life can be changed by using long-lasting material and altering design features such as area of cross section, types of fasteners used and so on. This rule may not necessarily apply to products in the fashion industry. That industry experiences constantly changing trends because of which this rule is rendered obsolete.

Rule 7: Designers should try and incorporate product features that reduce the need for maintenance and enhance product life. This can be achieved by using better, long-lasting materials, surface treatments or structural arrangements that protect from dirt, wear and corrosion.

Consider the example of a mini-chopper. Using stainless steel instead of steel would protect parts from corrosion. Sensitive parts and components should be protected using thermoplastic covers provided it does not interfere with functionality.

Rule 8: Products should be designed for ease of upgrading, repair and recycling (these topics will be dealt with in detail in future chapters). This can be achieved by incorporating design practices such as ease of accessibility, adopting modular product design, labeling and so on. Products should be designed such that they are easy to service. This enhances their maintainability.

Rule 9: The use of blended materials and alloys should be avoided. This promotes ease of upgrading, repair and recycling of the components. As far as possible, only a few, simple and recycled materials should be used if functionally acceptable.

Rule 10: The number of joining elements should be minimized as much as possible. On the basis of EoL option (whether the product is destined to be reused, recycled or remanufactured), components can be fastened using a variety of options such as screws, welds, adhesives, geometric locking, snap fits and so on. Instructions should also be provided for removing and recycling the toxic and precious material contained in the product.

1.9 CONCLUSION

This chapter had introduced readers to the basic concept of sustainability and the importance of sustainable product design in our contemporary world. To that end, readers were also given a succinct overview of the generalized product design process. An insight on how such a process can be potentially modified was offered through the ten golden rules of sustainable product design. It was realized that it is essential to "tweak" the generalized product design process in order to design sustainable products. This could be achieved by carefully selecting materials and processes that are able to accomplish the objective. Also, products that are designed not only for ease of manufacturability and assembly but also for ease of disassembly and maintenance are likely to last longer and are likely to be easier to recycle and remanufacture.

The topics of designing products for assembly and disassembly as well as recycle, reuse and remanufacture will be dealt with more thoroughly in the following chapters of the book.

REFERENCES

Alonso, R. (1990). *Continuous Improvement in Reliability*, pp: 106–114.

Anderson, D.M. (1991). *Design for Manufacturability*. CIM Press, Lafayette, CA.

Armstrong, L. and White, W. (1997). "Case study: Chrysler corporation life cycle management comparison of three engine oil filters," International Business Communications Environmental Cost Accounting, Conference Proceedings, Washington, DC.

Bieda, J. (1992). *Reliability Growth Test Management in the Automotive Component Industry*. pp. 387–393.

Blumberg, J., Korsvold, A. and Blum, G. (1997). "Environmental performance and shareholder value," *World Business Council for Sustainable Development*.

Caldecote, V. (1989). "Investment in new product development," in: Roy, R. and Wield, D. (eds.), *Product Design and Technical Innovation*. Open University Press, Milton Keynes, UK.

Comizzoli, R.B., Landwehr, J.M. and Sinclair, J.D. (1990). "Robust materials and processes: key to reliability." *AT&T Technical Journal*, Nov/Dec, pp: 113–128.

Consumer Reports (1990). *Three Ways to Brown Bread*. June, pp: 380–383.

Feldman, S.J., Soyka, P.A. and Ameer, P. (1997). "Does improving a firm's environmental management system and environmental performance result in higher stock price?" *Journal of Investing*, January 1997.

Fiksel, J. (1997). "Practical issues in environmental performance evaluation," in Tibor, T. and Feldman, I. (eds.), *Implementing ISO 14000*, Irwin.

Fiksel, J., Mcdaniel, J. and Spitzley, D. (1998). "Measuring product sustainability," *The Journal of Sustainable Product Design*, July 1998.

Hall, A.D. (1968). *A Methodology for Systems Engineering*. Sixth Edition. Van Nostrand. Princeton, NJ.

Haubner, P.J. (1990). "Ergonomics in industrial product design." *Ergonomics*, Vol: 33, No: 4, pp: 477–485.

James, P. (1997). "The sustainability cycle: A new tool for product development and design," *Journal of Sustainable Product Design*, July 1997 No: 2.

Lehni, M. (1998). "WBCSD project on eco-efficiency metrics and reporting: State of play report," World Business Council for Sustainable Development, February 1998.

Luttropp, C. and Lagerstedt, J. (2006). "EcoDesign and the ten golden rules: Generic advice for merging environmental aspects into product development," *Journal of Cleaner Production,* Vol: 14, pp: 1396–1408.

Nasser, S.M. and Souder, W.E. (1989). "An interactive knowledge based system for forecasting new product reliability." *Computers and Engineering*, Vol:17, No: 1–4, pp: 323–326.

Nielsen, J. (1992). *Usability Engineering*. Academic Press, Inc., San Diego, CA.

Nielsen, J. (1993). "Iterative user interface design." *Computer*, November, pp: 32–41.

Pye, D. (1989). "The nature of Design," in Roy, R. and Wield, D (eds.), *Product Design and Technical Innovation.* Open University Press, Milton Keynes, UK.

Ranganathan, J. (1998). "Sustainability rulers: Measuring corporate, environmental and social performance," World Resources Institute-Sustainable Enterprise Initiative, May 1998.

Umeda, Y. and Tomiyama, T. (1997). "Functional reasoning in design." *IEEE Expert*, March–April, pp: 42–48.

2 Environmental Legislations

2.1 INTRODUCTION

As we have studied in the first chapter of the book, environmental legislations are often the primary driver of environment conscious design decisions on the part of the product development team. Environmental legislations are the cumulative result of increasing awareness on the part of consumers and the public at large. This awareness is with respect to environmental problems such as pollution, climate change, greenhouse effect and so on. In general, people are beginning to pay increased attention to such issues and demand a better life for themselves and their children. The result is that lawmakers are now investigating the problem of environmental pollution in some depth. They are researching the role of contaminants that are responsible for causing the pollution. The goal is to pass legislation that will seek to alleviate the negative impact of such substances and conserve the environment. Environmental conservation is, after all, the first pillar of sustainability.

For instance, the general assembly of the state of Illinois made the following discoveries which in turn prompted them to pass environmental legislation in that state (www.ilga.gov/legislation/publicacts/95/PDF/095-0959.pdf).

- Electronic products are the fastest growing portion of the solid waste stream. In 2005, 2,600,000 tons of electronic products became obsolete yet only 13% of those products were recycled.
- Many electronic products contain lead, mercury, cadmium, hexavalent chromium, and other materials that pose environmental and health risks that must be managed.
- Many obsolete electronic products can be recycled or refurbished for reuse and then returned to the economic mainstream in the form of raw materials or products.
- Electronic products contain metals, plastics, and leaded glass that have resale value. The reuse of these components conserves natural resources and energy, and the reuse also reduces air and water pollution and greenhouse gas emissions.

- Management is necessary to place the reuse and recycling of obsolete residential electronic products as the preferred management strategy over incineration and landfill disposal.
- The Illinois Recycling Economic Information Study of 2001 estimated that the total economic impact of establishing statewide recycling and reuse programs for residential electronic products may result in the creation of nearly 4,000 new jobs and $740 million in annual receipts.

In the European Union (EU), the following discoveries were made (https://ec.europa.eu/commission/presscorner/detail/en/MEMO_05_248). Note that the EU reached some of the very same conclusions that were reached by the state of Illinois legislators, especially with respect to toxic materials such as lead, mercury, cadmium and hexavalent chromium present in electronic waste. The following material is taken from a memo released by the EU.

- Electrical and electronic waste is waste that can be traced to a large variety of products. Sources include IT and telecommunications equipment, lighting equipment, and consumer goods such as radios, small and large household appliances, TV sets, video cameras and hi-fi systems. This equipment is made up of many different materials and components, some of which are hazardous. Therefore, electrical and electronic waste can cause major environmental problems during the waste management phase, in particular landfilling and incineration, if it is not properly treated.
- As a matter of fact each electrical and electronic piece of equipment consists of a combination of several basic building blocks, such as circuit boards/assemblies, cables, cords and wires, plastics containing flame retardants, mercury switches, display equipment such as cathode ray tubes and crystal liquid displays, accumulators and batteries, light generating devices, capacitors, etc.
- Environmentally problematic substances in these components include certain heavy metals (mercury, lead, cadmium and chromium) and halogenated substances (CFCs, PCBs, PVCs and brominated flame retardants). Many of these substances can be toxic and may pose risks to human health when released. For instance, lead can damage the nervous system and can adversely affect the cardiovascular system and the kidneys. Cadmium also affects the functioning of kidneys and can cause brain damage.
- Up to now more than 90% of electrical and electronic waste is landfilled, incinerated or recovered without any pre-treatment, which means that the pollutants could be released into the environment and contaminate air, water and soil.
- Key data on electrical and electronic waste (WEEE) from 1998 showed a generation of 14 kg per inhabitant and year – in total, around 6 million tonnes per year (4% of the municipal waste stream). WEEE was estimated to be growing at 3–5% per year, which makes it the fastest growing waste stream, growing three times faster than the average waste stream. Today, citizens are likely to generate between 17 and 20 kg per head and year.

Through their observations, lawmakers make a very important point: in order to live a healthier life, it is necessary to properly manage electronic and industrial waste. Thus, it is necessary for companies to take the onus for not only designing and developing environmentally friendly products, but to also organize and manage product takeback programs with the goal of recycling, reusing or remanufacturing those products. Every ounce of waste that can be saved from being disposed of in a landfill goes a long way towards building a healthier and cleaner world, one step at a time. These standards have been drafted to obligate industry to continue assuming responsibility for their products even after their useful life. Products will be expected to be not only manufacturable and reliable but also serviceable and recyclable. They will be expected to derive minimal energy and resources from the environment and discharge minimal amount of wastes during and after their life cycles.

This chapter seeks to present some examples of environmental legislations passed in various states across the US, Germany, France and other countries such as China and India among others. It will also present some of the proactive initiatives taken by various electronic companies and manufacturing companies to minimize the impact of their products on the environment.

2.2 EXAMPLES OF SOME ENVIRONMENTAL LEGISLATIONS AROUND THE WORLD

This section of the chapter will present examples of legislations from around the world that deal with managing electronic waste and product takeback. It will also include specific language in terms of manufacturer responsibilities where pertinent. Let us start with examples of legislations passed by some US states.

2.2.1 Environmental Legislation for the State of New York

(www.dec.ny.gov/docs/materials_minerals_pdf/ewastelaw2.pdf):

Retailer requirements for the product takeback bill for the state of New York are specified below:

1 At the location of sale of covered electronic equipment, a retailer shall provide purchasers of covered electronic equipment with information, if any, about opportunities for the return of electronic waste that has been provided to the retailer by a manufacturer.

2 Beginning April first, two thousand eleven, no retailer shall sell or offer for sale in the state any covered electronic equipment unless the manufacturer and the manufacturer's brands are registered with the department pursuant to section 27–2605 of this title. If the retailer purchased covered electronic equipment from a manufacturer who fails to register by January first, 2011, or prior to the date the manufacturer withdrew its registration or the registration was revoked by the department, the retailer may continue to sell the covered electronic equipment for one hundred eighty days after April first, 2011, or the date the registration was withdrawn or revoked.

The state of New York imposed a ban on disposing of electronic waste into landfills. This disposal ban for the state of New York is delineated in the following paragraph:

1 Beginning April first, 2011, no manufacturer, retailer, or owner or operator of an electronic waste collection site, electronic waste consolidation facility or electronic waste recycling facility in the state shall dispose of electronic waste at a solid waste management facility or hazardous waste management facility, or place electronic waste for collection which is intended for disposal at a solid waste management facility or hazardous waste management facility.
2 Beginning January first, 2012, no person except for an individual or household shall place or dispose of any electronic waste in any solid waste management facility or place electronic waste for collection which is intended for disposal at a solid waste management facility or hazardous waste management facility in this state. Persons engaged in the collection of solid waste for delivery to a solid waste management facility shall provide written information to users of such facility on the proper methods for the recycling of electronic waste.
3 Beginning January first, 2015, no individual or household shall place or dispose of any electronic waste in any solid waste management facility or place electronic waste for collection which is intended for disposal at a solid waste management facility or hazardous waste management facility in this state.
4 Beginning January first, 2012, an owner or operator of a solid waste management facility or hazardous waste management facility shall educate users of such facility on the proper methods for the management of electronic waste. Such education shall include: (a) providing written information to users of such facility on the proper methods for recycling of electronic waste; and (b) posting, in conspicuous locations at such facility, signs stating that electronic waste may not be disposed of at the facility.

2.2.2 Environmental Legislation for the State of Illinois

(www.ilga.gov/legislation/publicacts/95/PDF/095-0959.pdf):

The consumer electronics recycling act of the state of Illinois aims to facilitate product takeback by adopting a variety of measures. In order to accomplish this goal, certain terminologies are set forth in this legislation. Examples are presented as follows:

- "Covered electronic device category" or "CED category" means each of the following 8 categories of residential CEDs:
  1 Computers and small-scale servers;
  2 Computer monitors;
  3 Televisions;
  4 Printers, facsimile machines, and scanners;
  5 Digital video disc players, digital video disc recorders, and videocassette recorders;
  6 Video game consoles;
  7 Digital converter boxes, cable receivers, and satellite receivers; and
  8 Electronic keyboards, electronic mice, and portable digital music players.

- "Manufacturer e-waste program": Any program established, financed, and operated by a manufacturer, individually or collectively as part of a manufacturer clearinghouse, to transport and subsequently recycle, in accordance with requirements of this Act.
- "Collector" means a person who collects residential CEDs at any program collection site or one-day collection event and prepares them for transport.
- "Recycling" includes any process by which residential CEDs that would otherwise be disposed of or discarded are collected, separated, or processed and returned to the economic mainstream in the form of raw materials or products.
- Recycler Responsibilities:
  1 Recyclers must comply with federal, State, and local laws and regulations, including federal and State minimum wage laws.
  2 Recyclers must implement the appropriate measures to safeguard occupational and environmental health and safety, through:
     a environmental health and safety training of personnel;
     b an up-to-date, written plan for the identification and management of hazardous materials;
     c an up-to-date, written plan for reporting and responding to exceptional pollutant releases.
  3 Recyclers must maintain on file proof of workers' compensation and employers' liability insurance.
  4 Recyclers must provide adequate assurance, such as bonds or corporate guarantees, to cover environmental and other costs of the closure of the recycler's facility, including cleanup of stockpiled equipment and materials.
  5 Recyclers must apply due diligence principles to the selection of facilities to which components and materials are sent for reuse and recycling.
  6 Recyclers must establish a documented environmental management system that is appropriate in level of detail and documentation to the scale and function of the facility.
  7 Recyclers must use appropriate equipment to properly process incoming materials and control releases to the environment. The dismantling operations and storage of residential CED components containing hazardous substances must be conducted indoors and over impervious floors. Storage areas must be adequate to hold all processed and unprocessed inventory. When heat is used to soften solder and when residential CED components are shredded, operations must be designed to control indoor and outdoor hazardous air emissions.

2.2.3 Environmental Legislation for the State of Virginia

State law for the state of Virginia has categorically spelled out retailer responsibilities in their product takeback program as follows (https://law.lis.virginia.gov/vacode/10.1–1425.31/):

> A person who is a retailer of computer equipment may not sell or offer to sell new computer equipment in the Commonwealth unless the equipment

is labeled with the manufacturer's label and the manufacturer has a recovery plan that complies with the provisions of this article and is accessible on the manufacturer's website. A retailer who is not a manufacturer is not required to collect computer equipment for recycling or reuse under this article.

Language pertaining to reporting requirements is included in the state law as follows (https://law.lis.virginia.gov/vacode/10.1–1425.30/):

- The name and contact information of the representative responsible for the manufacturer's recovery plan;
- The weight of computer equipment collected, recycled, and reused during the preceding calendar year;
- Documentation certifying that the collection, recycling, and reuse of computer equipment complies with § 10.1–1425.38 regarding sound environmental management.

The state law specifies requirements for manufacturers. They are as follows (https://law.lis.virginia.gov/vacode/10.1–1425.29/):

A Before a manufacturer may offer computer equipment for sale in the Commonwealth, the manufacturer shall:
  1 Adopt and implement a recovery plan; and
  2 Affix a permanent, readily visible label to the computer equipment with the manufacturer's brand.

B The recovery plan shall enable a consumer to recycle computer equipment without paying a separate fee at the time of recycling and shall include provisions for:
  1 The collection from a consumer of any computer equipment that has reached the end of its useful life and is labeled with the manufacturer's brand; and
  2 Recycling or reuse of computer equipment collected under subdivision 1.

C The collection of computer equipment provided under the recovery plan must be reasonably convenient and available to consumers in the Commonwealth and designed to meet the collection needs of consumers in the Commonwealth. Examples of collection methods that alone or combined meet the convenience requirements of this section include:
  1 A system by which the manufacturer or the manufacturer's designee offers the consumer a system for returning computer equipment by mail at no charge to the consumer;
  2 A system using a physical collection site that the manufacturer or the manufacturer's designee operates and to which the consumer may return computer equipment; and
  3 A system using collection events held by the manufacturer or the manufacturer's designee at which the consumer may return computer equipment.

D Collection services under this section may use existing collection and consolidation infrastructure for handling computer equipment and should encourage the inclusion of systems jointly managed by a group of manufacturers, electronic recyclers and repair shops, recyclers of other commodities, reuse organizations, not-for-profit corporations, retailers, recyclers, and other suitable operations. If a manufacturer or its designee offers a mail-back system as described in subdivision C 1, either individually, by working together with a group of manufacturers, or by working with others, it shall be deemed to meet the convenience requirements of this section.

E The recovery plan shall include information for the consumer on how and where to return the manufacturer's computer equipment. The manufacturer:
    1 Shall include collection, recycling, and reuse information on the manufacturer's publicly accessible website;
    2 Shall provide collection, recycling, and reuse information to the Department; and
    3 May include collection, recycling, and reuse information in the packaging or in other materials that accompany the manufacturer's computer equipment when the equipment is sold.

F Information about collection, recycling, and reuse on a manufacturer's publicly accessible website does not constitute a determination by the Department that the manufacturer's recovery plan or actual practices are in compliance with this article.

G If more than one person is a manufacturer of a certain brand of computer equipment as defined by § 10.1–1425.27, any of those persons may assume responsibility for and satisfy the obligations of a manufacturer under this article for that brand. If none of those persons assumes responsibility or satisfies the obligations of a manufacturer for the computer equipment of that brand, any of those persons may be considered to be the responsible manufacturer for purposes of this article.

H The obligations under this article of a manufacturer who manufactures or manufactured computer equipment, or sells or sold computer equipment manufactured by others, under a brand that was previously used by a different person in the manufacture of the computer equipment extends to all computer equipment bearing that brand regardless of its date of manufacture.

2.2.4 Environmental Legislations in Germany

The German environmental ministry was the first to put the issue on the political agenda and their regulation on end-of-life (EoL) vehicles can be considered to be the most stringent and lopsided of all European countries (Choi Athelstan, K.Y., 1994). The crux of the policy on combating waste from end-of-life vehicles (contained in the waste avoidance and waste management act [WMA] of November 1986) can be stated as follows:

- Producer responsibility for the processing of end-of-life vehicles.
- Increasing stringent norms on the final disposal of shredder residues.

In terms of greenhouse gas emissions and primary energy consumption, Germany has set ambitious goals for itself. For instance, the country has resolved to reduce its primary energy consumption by 20% by the year 2020 compared to 2008. This should be further reduced by 50% by 2050. In addition, greenhouse gas emissions should also be reduced by 40% compared to 1990 levels (Mckenna et al., 2013).

The German Environmental Ministry (BMU) through a series of draft policy papers laid down its policy on end-of-life vehicles. The first paper (Zielfestlegung: August 1990) obliged car manufacturers and importers to take back end-of-life vehicles without any cost incurred by the last owner. It required the reutilization of parts and materials if technically possible. It also required manufacturers or importers or a third party on their behalf to establish a dense network of car dismantling companies to comply with these obligations. Finally, and most importantly, it urged car manufacturers to consider the need for waste prevention in the conception of new vehicles.

Through this policy, the BMU extended the responsibility of the producer for the environmental impact of its products to encompass the products' entire environmental life cycle. The second draft (Altautoverordnung: August 1992) tended to concentrate more on new vehicle development and construction, including an extension of product durability. Automobile waste management (AWM) is rephrased in the second draft of the Altautoverordnung (1994). It defined AWM in terms of a closed loop economy. The basic difference between the first and second draft of the Altautoverordnung consists in the BMU agreeing to incineration as being a form of contribution to waste management, and in the second draft only under certain conditions. Specific recycling targets have been mentioned which have to be met through recycling of parts and materials as well as incineration with valorization of the heat generated.

Table 2.1 depicts recycling targets set by BMU for 1996 and 2000.

The salient features of the second draft of the Altautoverordnung are enumerated as follows:

- Manufacturers are required to produce automobiles with increased product durability, with a high degree of reusability, repeatability (ability to be replaced) and ease of repair.
- Material recycling is to be facilitated by material identification and selection of recyclable material in vehicle construction.

TABLE 2.1
Material recycling targets set by BMU in Germany

Material	Recycling targets for 1996 (%)	Recycling targets for 2000 (%)
Steel	100	100
Non-ferrous metals	85	90
Plastics	20	50
Rubbers	40	50
Elastomers	20	30
Glass	30	50

Source: Choi Athelstan (1994).

- Car manufacturers are required to take back free of charge end-of-life vehicles from the last owner.
- Car manufacturers are obligated to set up a nationwide network of return stations with the same density as the dealer network.
- The responsibility of the consumer is also underlined in the second draft of the Altautoverordnung. The last owner of the vehicle is required to deliver his/her end-of-life vehicle at a licensed collection point. Upon doing so, the owner is given an official certificate of disposal relieving him of tax obligations.

A new waste management act came into existence in Germany in 1996 with the advent of the "Closed Substance Cycle and Waste Management Act" (Feldmann et al., 1999). The new law states the following principles of waste treatment:

- The avoidance of waste has top priority.
- Unavoidable waste must be recycled.
- Unrecyclable waste needs to be disposed of without causing any harm either to the environment or to human health.

Since the end user of a product has only a minimal influence on its recyclability, the law assigns explicit responsibility to the producers for their products. According to an ordinance passed in 1998, automobile manufacturers are obligated to take back their cars and recycle them at a rate of 85% until 2002 and at 95% till 2015.

2.2.5 Environmental Legislations in France

The French approach can be considered to be more liberal and democratic than its German counterpart since it entails mutual cooperation between all parties involved. The French regulatory process was initiated by the European Commission. The strategy applies equally to products and processes and aims to close material flows into loops. The policy of the European Commission pertaining to waste prevention constitutes a very proactive approach by explicitly calling for marketing of products that have been developed to minimize waste (in quantity and toxicity) during all stages of the products' life cycle. Industry offered to participate in the decision-making process so as to prevent direct regulation. The group (accord cadre) reached the following agreement as concerns various targets in March 1993:

- By the year 2002, a maximum of 15% waste disposal per car, a maximum of 200kg per vehicle.
- For new models (2002 onwards), a maximum of 10% waste disposal.
- A maximum of 5% waste disposal in the long run.

An important clause in the agreement states that operators (dismantlers, shredder companies, metal recovery firms, etc.) may freely choose from among the different technologies and various modes of valorization available. The coordination of different activities to be undertaken by each of the parties is supposed to take place through a free market system. This includes prices to be paid in various transactions in vehicle collection, pretreatment, disassembly, recovery and recycling as well as the conditions under which the last owner's vehicle is to be accepted by the operators. The major advantage of the involvement of all parties in the decision-making process is mutually acceptable targets, more technical and economic feasibility and shared responsibility.

2.2.6 Environmental Legislations in England

The following is a chronology of the development of the British policy:

- Early initiatives are taken by CEST: a private, nonprofit consultant to design a strategic approach for analyzing and solving the end-of-vehicle waste problem. The CEST group involved both the government and industry (1991).
- The CEST initiative passes on to the Acord (Automotive Consortium on Recycling and Disposal) group (December 1991) established by the SMMT (the Society of Motor Manufacturers and Traders). The members of the Acord group represent manufacturers, importers, vehicle disposal industries, trade associations and the government.
- The end of life vehicle disposal (Acord, 1992) and
- The preliminary operating plan (Acord, 1993) is mutually agreed upon. The objectives and terms are closely modeled on the French accord cadre.

2.2.7 Environmental Legislations in the EU

The following objectives for reducing end-of-life vehicle waste are proposed (ELV Project group, 1994: Annex C):

- At present, 75% of all end-of-life vehicles are collected for recovery of metal content.
- By 1998, all end-of-life vehicles should be collected, drained and cleaned.
- By 2002, at least 85% of the initial weight of the end-of-life vehicles should be recycled.
- By 2002, at least 90% of the initial weight of the end-of-life vehicles from new models should be recycled.
- By 2015, at least 95% of the initial weight of end-of-life vehicles should be recycled.

2.2.8 Environmental Legislations in the Netherlands

The Dutch approach is a combination of the German and French approaches. The environmental ministry defined specific recycling targets as an input to the regulatory process. The strategic discussion process was structured precisely on the German waste management policy. The application of the strategic discussion process resembles the French approach. The environmental ministry was more open to discuss various means through which targets could be realized.

2.2.9 Waste Electrical and Electronic Equipment (WEEE) Directive in the EU

The goal of the EU directive on waste electrical and electronic equipment (WEEE directive) is to prevent the generation of electrical and electronic waste and to promote reuse, recycling and other forms of recovery in order to reduce the quantity of such waste to be eliminated through landfilling or incineration. The directive requires the collection of WEEE, recovery and reuse/recycling. As appropriate, priority should be given to reusing the whole appliance.

As compared to end-of-life aspects of automotive products, the end-of-life processing of WEEE is still greatly lacking in efficiency and effectiveness (Nagel and Meyer, 1999). German information technology association CYCLE or its European counterpart EUROBIT has been heavily involved in discussions with the legislative participants in order to achieve a consensus policy on the taking back and recycling of WEEE (Nagel and Meyer, 1999). In addition, research initiatives are trying to foster integrated and comprehensive approaches towards suitable and sustainable end-of-life solutions. WEEE was part of the EU priority waste stream program in the early 1990s, primarily due to its volume and hazardous potential. The total volume of WEEE has been estimated to be around 6.5–7.5 million tons per year within 19 EU alone. This is equivalent to 1% of the EU's solid waste and 4% of the municipal solid waste (Nagel and Meyer, 1999). Several EU member states have been actively working on forming legislation to combat the menace of WEEE. In Germany, the first draft of the "Electronic Scrap Ordinance" was published in 1992. It covered the entire range of WEEE. Despite many years of intense discussions between various stakeholders, an all-encompassing solution that satisfies everyone continues to be elusive.

The following recommendations have been suggested in order to minimize WEEE. These objectives are expected to be reached through several regulations:

- Ban on heavy metals such as mercury, cadmium, and so on.
- Takeback obligation for the manufacturer and free of charge for the final user.

Table 2.2 indicates the collection, reuse and recycling targets for WEEE. The directive was expected to be due for discussions at the national level by the year 2000.

The Dutch government plans to reduce the amount of electronic waste and dispose of the remaining waste without any leakages to the environment. These objectives are realized by 21 developing and implementing courses of action and tools that will enable manufacturers to practice waste control. The results that the Dutch government is looking for are the following (Ron Ad de and Penev, 1995):

- Avoidance of the creation of waste during the production and the consumption process.
- Reusage of unavoidable waste as far as possible. This could be achieved perhaps after reprocessing or using it in the same or other production or consumption chains.
- Removal of avoidable waste with negligible risk to the environment.

TABLE 2.2
Collection, reuse and recycling targets for WEEE

Type of equipment: List products separately	Minimum reuse and recycling targets (%)
Large household appliances	90
Small household appliances	70
IT equipment	70
Telecommunications equipment	70
Radio, TV, electroacoustic and musical equipment	70
Medical equipment systems	70
Monitoring and control instruments (except industrial process control equipment)	70
Toys	70
Electrical and electronic tools	70
Gas discharge lamps	90
CRT containing equipment	90
CFC, HCFC, HFC containing equipment	90

Source: Nagel (1999).

In December 2008, it was proposed that the WEEE directive be revised. It was revised and entered into force on August 13, 2012. The following are some salient features contained in the new directive especially with respect to preferring reusing products over recycling them (https://eur-lex.europa.eu/legal-content/EN/TXT/?uri=CELEX:32012L0019):

- The collection, storage, transport, treatment and recycling of WEEE as well as its preparation for re-use shall be conducted with an approach geared to protecting the environment and human health and preserving raw materials and shall aim at recycling valuable resources contained in EEE with a view to ensuring a better supply of commodities within the Union.
- Where appropriate, priority should be given to preparing for re-use of WEEE and its components, subassemblies and consumables. Where this is not preferable, all WEEE collected separately should be sent for recovery, in the course of which a high level of recycling and recovery should be achieved. In addition, producers should be encouraged to integrate recycled material in new equipment.
- The recovery, preparation for re-use and recycling of WEEE should be counted towards the achievement of the targets laid down in this Directive only if that recovery, preparation for re-use or recycling does not conflict with other Union or national legislation applicable to the equipment. Ensuring proper preparation for re-use, recycling and recovery of WEEE is important for sound resource management and will optimize supply of resources.

2.2.10 ROHS Directive

ROHS is an acronym for "Restriction of Hazardous Substances." It has a direct impact on the entire electronics industry as well as many electrical products. The directive originated in the EU in 2002. Its focus is on restricting six substances which are hazardous in nature and are commonly found in electronic devices. It is mandated that beginning July 1, 2006, all applicable products in the EU must pass ROHS compliance. The ROHS 3 directive added four more substances to the original list. ROHS 3 is effective beginning July 22, 2019.

In order to be ROHS compliant in the year 2020, a business that sells electrical or electronic products, equipment, cables, subassemblies, components or spare parts to ROHS countries or to any other entity such as resellers, distributors or integrators who in turn sell products to such countries is impacted if they use any of the ten listed substances.

The EU ROHS specifies maximum limits for the following substances, which are restricted (ppm: Parts per million):

- Cadmium < 100 ppm
- Lead < 1,000 ppm
- Mercury < 1,000 ppm
- Hexavalent Chromium < 1,000 ppm
- Polybrominated Biphenyls < 1,000 ppm
- Polybrominated Diphenyl Ethers < 1,000 ppm
- Bis (2 -Ethylhexyl) phthalate < 1,000 ppm
- Benzyl butyl phthalate < 1,000 ppm
- Dibutyl phthalate < 1,000 ppm
- Diisobutyl phthalate < 1,000 ppm.

2.2.11 R2 Standard

The R2 standard deals with safely disposing waste electronics. It was developed and is certified by SERI which is the housing body and ANSI-accredited standards development organization for the same.

The R2 standard exists due to a collaborative partnership between public, private and nonprofit sectors. R2 stands for "Responsible Recycling" practices for use in accredited certification programs. It is the results of the US Environment Protection Agency's (EPA) push to convene a multistakeholder process to create a voluntary, market-based mechanism to ensure best practices in electronic recycling (sustainableelectronics.org).

The standard was developed as the result of a transparent, balanced and consensus-based process. Different stakeholders were included in the group including representatives from the following:

- US EPA.
- Regulators from various state agencies.

- Electronics recyclers, refurbishers and representatives from their respective trade organizations.
- Original equipment manufacturers (OEMs) and customers of electronics recycling services.
- Non-government organizations.

Electronics recyclers can gain R2 certification. Certified electronics recyclers have demonstrated through audits and otherwise that they continually meet specific high environmental standards. They have also demonstrated that they can safely manage used electronics. The benefits of gaining R2 certification are as follows:

- Reduction of environmental and human health impact occurring as a result of improper recycling.
- Providing increased access to quality reusable and refurbished equipment to potential customers.
- Reduction of energy use and environmental impact associated with mining and processing of virgin materials, thereby conserving natural resources that are limited (epa.gov).

2.2.12 E Stewards Standard

The e stewards standard is a company-level certification based on a standard that was developed by the US EPA and industry partners. The certification applies to electronics recyclers, asset recovery companies, refiners, processors and other entities in the electronics industry. The standard requires the following (estewards.org):

- Secure data: All stored data from electronic devices must be destroyed.
- Best recycling practices: Toxic electronic waste should be managed in keeping with the best available practices in order to protect workers and the environment.
- Ethical labor: All child labor, coerced labor and prison labor are banned.
- Toxic materials should be responsibly managed downstream: Recyclers that are e stewards certified should track all toxic materials downstream and ensure that they are properly managed.
- Legal and responsible exports: Based on international law (Basel Convention), export of toxic electronic waste to developing countries is not allowed.
- Organization-wide conformance: All facilities and operations belonging to the company in each country should be in conformance with the certification.

2.2.13 Environmental Laws in China

BBC reports as of July 2019 that Shanghai has become the first Chinese city to make it mandatory for residents to sort and recycle household trash (www.bbc.com/news/world-asia-china-48847062). Noncompliance is likely to incur heavy fines and detrimentally impact residents' credit rating. The most populous city in the world with more than 24 million residents currently recycles about 10% of its waste. According to official statistics, 3,300 tons of recyclables are collected every day. According to

Xinhua News Agency, Shanghai produces nine million tons of trash every year. The new legislation requires trash to be divided into the following four categories:

- Recyclable goods such as cans and bottles.
- Harmful waste such as batteries and medicine.
- Kitchen waste including food.
- Other waste.

Entities including individuals and organizations who are found to not be in compliance with this law will be subjected to substantial fines. For example, non-complying individuals can be fined up to 200 yuan whereas organizations could be fined up to 50,000 yuan.

2.2.14 Product Takeback Law in India

The ministry of environment and forest placed legal liability for reducing and recycling electronic waste with producers under the so called e-waste rules in 2011. The rules came into effect beginning May 1, 2012 and are part of the Environment Protection Act.

These rules require that manufacturers and importers of computers, mobile phones and "white goods" organize e-waste collection centers and product takeback systems. These rules are applicable to every producer, consumer and bulk consumer that are involved in manufacture, sale, purchase and processing of electronic equipment or components.

As of 2011, India had produced as much as 400,000 tons of electronic waste every year of which 19,000 tons had been recycled.

According to the new rules, producers will be required to provide consumers with information on equipment disposal after use. This is meant to prevent consumers from disposing of electronic waste together with other domestic waste. Producers are also required to make the consumer aware of the hazardous components and materials present in the electronic equipment.

As far as commercial consumers and government departments are concerned, they will be required to recycle their e-waste by channeling it to recycling centers or making sure the waste is taken back by the suppliers.

2.2.15 Product Takeback Program: Caterpillar Reman

Caterpillar, a manufacturer of engines and construction equipment such as earthmovers, is at the forefront of environmentally friendly manufacturing practices such as recycling and remanufacturing. The emphasis is on keeping resources within the value chain by focusing on the circular flow of materials, energy and water. This optimizes the use of resources and maximizes the total life cycle value of the company's products. Doing so minimizes the cost of ownership for customers. The company's thrust into the realm of remanufacturing is through its so-called Cat Reman program. It provides customers with lower cost products, shorter downtimes and quick, dependable service options. The company aims to achieve a zero-landfill footprint goal by the year 2020 (www.caterpillar.com/en/company/sustainability/remanufacturing.html).

The Cat Reman program returns products that are at the end of their lives to "same as new" condition. We will take an in-depth look at the concept of remanufacturing later in this book. According to the company, doing so reduces the customer's owning and operating costs by providing them with a part that is good as new at a fraction of the cost of the new part. The remanufacturing process enables the company to reduce waste, lower the production of greenhouse gases and minimize the need for raw materials. According to the company, reusing components helps them use energy and materials more efficiently (Caterpillar sustainability remanufacturing).

The remanufacturing program is based on an exchange system. A used component is returned to the company by customers. In return, a remanufactured replacement component with the same as new quality, reliability and warranty are received. Examples of rebuild programs run by the company include Cat-certified rebuilds, component overhauls at Cat dealers, solar turbine rebuilds and progress rail service rebuilds.

The Cat-certified rebuild consists of the following:

- More than 350 tests and inspections.
- Automatic replacement of 7,000 parts.
- A like new machine warranty.
- Service by trained dealer professionals using genuine equipment and parts.

Customers can maximize the value of their built-in equipment by resorting to remanufacturing and rebuild programs run by Caterpillar. Maximization of value occurs as a result of the following:

- Reliability and equipment uptime are increased.
- Cost-effective performance is ensured.
- A like new warranty is received.
- Customer is provided with a variety of repair options whereby their service needs can be met.
- Customer is provided with a higher resale value.
- Majority of energy and materials required to make the original product/machine are conserved.
- Productivity is maximized.

2.2.16 Product Takeback Program: Dell

Dell Technologies Inc. is an example of an electronics company that is pioneering the sustainability movement in the electronics industry. According to their website, the company offers a host of options for households and businesses to recycle shipping material. The company also runs a variety of programs to enable product takeback of the many products that it sells to consumers.

According to Louise Koch, corporate sustainability director, the company achieved a savings of almost $2 million over the past five years as a result of using closed loop

recycled plastics in the manufacture of its many products. The company is aiming to expand this program (Bloomberg NEF). Since the program began five years ago, about 10,000 metric tons of closed loop plastics have been used in the company's products (such as desktops and tablets). The company offers a takeback system for its products for customers in more than 75 countries to help wipe data and dispose of their waste responsibly.

In order to contribute to the cost of public collection, the company pays a fee for every product that is sold. This is in keeping with EU law. As part of the company's takeback program, the customer receives a report on where their product eventually went. According to Dell, the company was able to take back approximately 900,000 metric tons of electronic waste by 2020.

2.2.17 Product Takeback Program: Ford

The Ford core recovery program (FCRP) was initiated in 2003. The program requires dealers to send back specific parts that are damaged or worn so that they can be recycled or remanufactured. The parts are restored to exacting standards as a result of the remanufacturing process. According to the company, the FCRP process recycles or remanufactures items such as plastic headlamps, taillamps, engine components and other steel and aluminum materials (Ford Recycling). This process prevents waste from being disposed of in landfills (www.fenderbender.com/articles/3513-core-charge-programs-ramping-up).

Over the course of its existence, the company estimates, the process had been able to successfully keep a huge amount of material from ending up in landfills. Specifically, this includes 149 million pounds of vehicle parts. Also, since 2011, the company had recycled more than 272,000 headlamps and 379,000 bumper covers.

The FCRP process consists of overseeing the collection, recycling and remanufacturing of damaged parts (such as the ones described). It also includes small sensors and large engine parts from Ford vehicles that have been repaired through the company's dealer network.

2.2.18 Product Takeback Program: Daimler

According to Daimler Corporation, the manufacturer of Mercedes Benz vehicles, more than 95% of a vehicle can be recovered (www.daimler.com/sustainability/resources/from-old-to-new.html). According to the company, its goal is to minimize not only emissions but also the consumption of resources over the lifespan of a vehicle. This is because the resources are valuable and are available only in limited amounts. Resource consumption over the lifecycle of a vehicle is minimized as a result of employees from production and development sectors thinking about the vehicle's "second life" from the very beginning. As a result, fewer automotive parts are disposed of in landfills thanks to clever product dismantling and recycling strategies (which happens to be the topic of this book). Parts that are still functional at the end of a vehicle's useful life are removed and reused in a new vehicle.

The Mercedes' Used Parts Center (UPC) is located in Neuhausen near Stuttgart. Every year, the UPC team carefully inspects more than 5,000 vehicles that are at the end of their useful lives so that as many useful parts as possible can be reused and sold. According to the company, "a used part that lives on in another Mercedes Benz vehicle attains the highest level of recycling (Daimler.com/sustainability)."

The recycling practiced at UPC is not limited to just parts. The center also tries to recover all of the operating materials. This includes brake fluid and coolant and other valuable resources such as glass, aluminum and copper. Parts that do not meet the company's high-quality standards are sent to the materials separation unit. According to EU stipulation, 95% of an end-of-life vehicle's material in weight needs to be recovered. Daimler believes they have actually surpassed this goal in some instances.

In this chapter, we have familiarized ourselves with a variety of environmental legislations from around the US and the world. We have also gained familiarity with some landmark product takeback practices in industry. Table 2.3 presents a comparison between various electronics takeback legislations from the different US states. Table 2.4 depicts what types of products are covered under which state legislation.

2.3 KEY TAKEAWAYS FROM STATE LAWS DEALING WITH DISPOSAL OF ELECTRONIC WASTE

As we have studied in this chapter so far, as many as 25 US states have passed laws mandating recycling programs so they can better deal with electronic waste. Of these, 23 states have incorporated some degree of the so called producer responsibility in their laws. This implies that manufacturers are financially responsible for recycling their products. Thus, it is mandatory that manufacturers bear the economic burden of recycling their products at the time of their useful lives. However, this "environmental conscious" attitude seems to be working differently in different states.

For instance, as much as 24, 370, 894 lbs of total electronic waste was collected in the state of Texas in 2010 whereas in Oregon, this waste amounted to 24,149,774 lbs. It is clearly understood that the population of Texas is much higher than that of Oregon and yet the amount of waste generated in pounds in both states is approximately identical. This is further corroborated by the fact that in 2010, the amount of electronic waste per capita in Texas (at 0.97 lbs) was significantly higher than in the state of Oregon (6.31 lbs). What can possibly explain this condition? A critical analysis of product takeback and some of the procedures adopted to recycle e-waste is necessary in order to attempt to arrive at an explanation.

In this section, we will examine the key takeaways that are a result of comprehending state laws that deal with e-waste and studying trends in product takeback practices (www.electronicstakeback.com).

2.3.1 Takeaway 1

2.3.1.1 E-waste Collection Volume Depends on Collection Infrastructure and/or Clearly Documented Collection Goals

Some states have a very well-established e-waste collection infrastructure whereas other states do not. This infrastructure makes the collection of electronic waste

TABLE 2.3
State laws on electronic recycling at a glance

State	Date law was signed	Start date of program collection	Types of products covered	Who will get free recycling?	Cost is borne by whom?	What is the information on toxics?	Goals/targets for collection/ convenience metrics	Does it include ban on prison labor?	Does it include disposal ban?
Connecticut	July 6, 2007	July 1, 2009 Program began early 2011 upon recycler approval	• TVs • Monitors • Laptops • Printers • Personal Computers	• Consumers • Any resident who drops off 7 products or fewer at once	Return share Municipalities facilitate collection and transportation to recyclers Recyclers bill manufacturers Market share is used to • Bill TV Companies • Determine all registration fees. • Determine orphan product share (computer companies)	None	States to have established collection goals by October 2010 statewide	No	Yes, this is effective January 2011
Hawaii	July 2008. TVs were to be added in 2009	January 2, 2010	• Computers • Laptops • Printers • Monitors • TVs were added in 2011	• Consumers • Nonprofit organizations • Government • Small and large businesses	It is essential for manufacturers to establish plans to collect and recycle their products	No	No	No	No
Illinois	September 17, 2008 Bill was revised in 2011	January 1, 2010	Scope to determine manufacturers obligation: Covered electronic devices	Consumers	Overall goal statewide is return share goal This was increased up to 10% over the goal for the previous year	Companies must disclose whether their products comply with ROHS.	Collection goals: Statewide per capita goals are translated into individual manufacturer goals	Yes	Yes, beginning 2012

(continued)

TABLE 2.3
(Cont.)

State	Date law was signed	Start date of program collection	Types of products covered	Who will get free recycling?	Cost is borne by whom?	What is the information on toxics?	Goals/targets for collection/ convenience metrics	Does it include ban on prison labor?	Does it include disposal ban?
			• Computers • Laptops • TVs • Monitors • Printers, scanners and faxes • Keyboard, mouse • MP3 players, video game consoles • VCRs and DVD players • Mobile phones • Computer cables • PDA • Zip Drives		This goal is converted into company obligations based on TV companies' market share and IT companies' return share				
Indiana	May 13, 2009 Revised in 2012	April 1, 2010 Program year runs from April through March	Video display devices such as • TVs • Monitors • Laptops Other electronic devices such as • Desktops • Printers • Keyboards • Fax Machines • VCRs and DVD players	• Households • Public schools • Small businesses with less than 100 employees	Market share Manufacturer pay for collection, transportation and recycling and meet goals based on market share of video display devices sold	Companies are required to report on display devices sold that exceed the maximum ROHS level toxics	Collection Goals Year 1: Manufacturer must recycle up to 60% of what was sold by weight in the previous year Year 2 and beyond goal: 80% If goals are not reached, penalties begin in year 3	No	Yes, starts in 2011

Maine	2004 Revised in 2009 Revised again in 2011	January 2006	Initially, covered • TVs • Monitors • Laptops CPUs not covered unless they are attached to monitor. Added in 2009 • Printers • Digital Picture frames as monitors • Video game consoles	In 2004, households only Added in 2011 • Primary and secondary schools • Nonprofit organizations • Small businesses with less than or equal to 100 employees	Manufacturers pay for transport and recycling costs and some collection costs as well Municipalities pay for some collection costs IT companies split costs by return share TV companies split costs by market share as of 2010	No	No	No	No
Maryland	2005	January 2006	• Monitors • Computers CPUs • Laptops • TVs added in 2007	None specified	Manufacturer pays fees to state State fees reimburse counties who pay for recycling through grants	No	No	No	No
Michigan	December 26, 2008	April 1, 2010	• Computers • Monitors • TVs • Laptops • Printers added in April 2011	• Consumers • Small business that drops of 7 or fewer units per day	Manufacturer pays for collection, transportation and recycling No level of service mandated	No	TV companies have a nonbinding goal of 60% by weight of what company sold during the previous year	Yes	No (will be studied)
Minnesota	May 8, 2007 Revised in 2009 Revised in 2011	August 2007	Video display devices > 9 inches • TVs • Monitors including security monitors. • Laptops, tablets, e-readers, digital picture frames. • Portable DVD players. • Desktops • Peripherals	Consumers	Manufacturer pays for collection, transportation and recycling Market share goals	Companies must report on display devices sold to households if they exceed the maximum ROHS levels for lead, mercury, cadmium and hexavalent chromium	Year 1: Manufacturers must recycle amount equal to 60% of what they sold by weight during the prior year Year 2: 80% of prior years sales	Yes, except for profit refurbishment and reuse activities	Was already in place

(continued)

TABLE 2.3 (Cont.)

State	Date law was signed	Start date of program collection	Types of products covered	Who will get free recycling?	Cost is borne by whom?	What is the information on toxics?	Goals/targets for collection/ convenience metrics	Does it include ban on prison labor?	Does it include disposal ban?
			• FAX machines • DVD players • VCRs.						
Missouri	June 16, 2008	Collection begins after July 1, 2010	• Monitors • PCs • Laptops • No TVs	Consumers	Manufacturer pays for collection, transportation and recycling No level of service is mandated	No	No	No	No
New Jersey	January 15, 2008 Revision signed January 2009	January 1, 2011	• TVs • Monitors • PCs • Laptops	• Consumers • Small businesses with 50 or less employees	Return share Manufacturer pays for collection, transportation and recycling TV companies assign costs of collective return share via market share	Must comply with ROHS on heavy metals	Goals to be set by January 2011	Yes	Yes, as of January 1, 2011
New York City Has been preempted by NY state law	April 1, 2008	7/1/2009 Preempted by state law	• TVs • PCs • Monitors • Laptops • Printers • Keyboard • Mice	Everyone—consumers, businesses etc.	Market share Producers must collect and recycle	No	Yes Collection goals based on market share 2012: 25% 2015: 45% 2018: 65%	No	Yes, as of July 1, 2010

New York State	May 29, 2010	April 1, 2011	• PCs, laptops, tablets, e-readers • TVs • Smallscale servers • Computer peripherals • VCR, DVRs, DVD players, portable digital music players, satellite convertors, video game consoles	• Individuals • Nonprofits < 75 employees • 501©3 nonprofits. • Schools • Government agencies • Small businesses < 50 employees Large nonprofits and businesses may be charged	Producers pay for collection, transportation and recycling based on their market share. Law establishes a statewide goal, then producers are assigned their portion based on their market share. Producers must also take back one unit for every unit sold.	Yes, They must disclose any products for sale that do not comply with ROHS.	Yes, statewide collection goals per person 2011: 3 lbs 2012: 4 lbs 2013: 5 lbs After 2013, goal is recalculated based on experience	No	Yes, as of April 1, 2011 for producer, retailers and waste handlers and effective January 1, 2015, for consumers
North Carolina	August 31, 2007	January 1, 2010 (2008 law delayed start till 2010)	2007 law • Desktops • Laptops • Monitors • Keyboards and mice 2008 law added • TVs and delayed start by 1 year 2010 law added • Printers • Scanners • Multifunction	2007: Nonspecified 2010 law clarified that it applies to • Consumers • Nonprofits of less than 10 employees	Producers must pay registration fees Computer manufacturers submit plans for how they will collect and recycle covered items Three tiers of plans with the least rigor (tier 1) requiring slightly higher fees Market share for TV companies Return share for IT companies	No	No	No	Yes, landfill and incinerator ban as of July 1, 2011
Oklahoma	May 13, 2008	January 1, 2009	• Desktops • Laptops • Monitors • NO TVs	Consumers	Producers pay for collection, transportation and recycling No level of service is mandated	No	No	No	No

(*continued*)

TABLE 2.3
(Cont.)

State	Date law was signed	Start date of program collection	Types of products covered	Who will get free recycling?	Cost is borne by whom?	What is the information on toxics?	Goals/targets for collection/ convenience metrics	Does it include ban on prison labor?	Does it include disposal ban?
Oregon	June 7, 2007 SB 82: Bill updated signed June 9, 2011	January 1, 2009	• TVs • Monitors • PCs • Laptops Beginning January 2015, printers, keyboards and mice were added	• Households • Small businesses • Small nonprofits • Anyone else including businesses dropping off seven items or less to collection points,	Producers pay for collection, transportation and recycling TV companies assign costs of collective return share via market share	No	Yes, convenience measures Producers must have collection site in every county, plus every city over 10,000	No	Yes, effective January 1, 2010
Pennsylvania	November 23, 2010	January 2012	• TVs • Monitors • Computers • Laptops • Peripherals	• Consumers • Small businesses with 50 or less employees	A manufacturer shall establish, conduct and manage a plan to collect, transport and recycle an amount of covered devices equal to the manufacturer's market share	No		Recyclers must be certified to R2 (allows prison) or e-Stewards no prison).	Yes, as of January 24, 2013
Rhode Island	June 27, 2008	February 1, 2009	• Computers • Laptops • Monitors • Televisions	Household or public and private elementary and secondary schools	Producers pay for collection, transportation and recycling	Video display devices sold that exceed ROHS levels must be disclosed	No	Yes	Yes, as of January 31, 2009
Texas	June 15, 2007 TV law June 2011	Sept 1, 2008	• Desktops • Monitors • Laptops • TVs added by 2011 law starting 2013	Consumers	Producers must have a program; no level of service is maintained	No	Not for computers Some target for TVs based on market share	No	No

South Carolina	May 19, 2010	July 1, 2011	• Computers • Laptops • TVs • Monitors • Printers	Consumers	Producers should have a takeback program; no level of service is mandated	No	No	No	As of July 1, 2011, covered products banned from landfill
Vermont	April 21, 2010. Revised 2012	July 1, 2011	• Computers • Laptops • TVs • Monitors • Printers • Peripherals	• Households • Charity • School district or small business with less than 11 employees. • Any resident dropping off 7 or fewer items per day (added for 2013).	Market share goals and convenience requirements are combined Essential to have 3 sites/county and an additional site in every city of 10,000 or more	No	Yes, per capita goals for collection and convenience requirements are set	No	Yes, as of January 1, 2011
Virginia	March 11, 2008	July 1, 2009	• Desktops • Laptops • Monitors • No TVs	Consumers	It is necessary that producers have a takeback program; no level of standard is mandated	No	No	No	No
Washington	March 2006	January 2009	• TVs • Monitors • Laptops • Desktop computers	Consumers, charities, small businesses, schools and small governments	Producers pay for collection, transportation and recycling Return share	No	Yes, convenience measures Producers must have collection site in every county plus every city over 10,000	Yes	This is not in the bill Some counties have passed bans
West Virginia	April 1, 2008	January 2009	• TVs • Monitors • Laptops • Desktop computers	Consumers	Producers pay a registration fee for $10,000 if they have no takeback program. They pay $3000 if they have a program.	No	No	No	No

(continued)

TABLE 2.3
(Cont.)

State	Date law was signed	Start date of program collection	Types of products covered	Who will get free recycling?	Cost is borne by whom?	What is the information on toxics?	Goals/targets for collection/ convenience metrics	Does it include ban on prison labor?	Does it include disposal ban?
Wisconsin	October 23, 2009	January 2010	<u>Scope for manufacturer obligation:</u> Video display devices such as TVs, monitors and laptops. Also, printers. <u>Scope for free collection:</u> • TVs • Monitors • Laptops • Desktops • Printers • Keyboards • DVS Players • Fax machines • VCRs.	Consumers (households)	Producers pay for collection, transportation and recycling based on their market share. The goal is 80% by weight of products sold to households and schools 3 years earlier.	Yes, producers must declare which products they sell that do and do not comply with ROHS directive	Yes, collection goals. Goal is 80% by weight of products sold to households and schools 3 years earlier.	Yes	Yes, as of September 1, 2010.
States that are not based on producer responsibility									
California	September 25, 2003	January 2005	TVs and monitors only. Portable DVDs added in 2006. Not CPUs and other products.	All owners-consumers and businesses.	Consumers pay a fee at purchase. Fee money goes to the state and is used to reimburse recyclers and collectors.	Comply with RoHS directive on heavy metals.	Bill set goal to eliminate electronic waste stockpiles and legacy devices by December 31, 2007	No	Was already in place.

					Companies cannot sell laptops, monitors, TVs, portable DVD players that exceed RoHS levels for Lead, Mercury, Cadmium and Hexavalent Chromium.			
Other models								
Utah	March 2011	No Collection required		Producers must report to the state by August 2011 and do public education by January 2012 about the recycling options available in the state. They are not required to do any recycling themselves.	No	No	No	No

Source: Adapted from Electronics Takeback Coalition. www.electronicstakeback.com.

TABLE 2.4
Scope of electronic waste covered by state laws

State	TVs	Monitors	Desktops	Laptops	Tablets	E-readers	Printers	Fax machines	Scanners	Keyboard mice	DVD players	VCRs	Convertor boxes	Receivers cable, sat	Cell phones	Game consoles	MP3 players	Small servers	Digital picture frames	GPS
California	**Xx**	**Xx**		**Xx**	**Xx**						**Xx***									
Connecticut	X	X	X	X	X	X	X													
Hawaii	X	X	X	X	X	X	X													
Illinois	X	X	X	X	X	X	X	X	X	X	X	X	X	X	**Xx**	X	X	X		
Indiana	X	X	**Xx**	X	X	X	**Xx**	**Xx**		**Xx**	**Xx**	**Xx**	**Xx**	**Xx**		**Xx**	**Xx**		**Xx**	**Xx**
Maine	X	X		X	X	X	X				X					X			X	
Maryland	X	X	X	X	X	X		X											X	
Michigan	X	X	X	X	X		X	X*	X*											
Minnesota	X	X	X	X	X	X	**Xx**	**Xx**		**Xx**	**Xx**	**Xx**							X	
Missouri		X	X	X																
New Jersey	X	X	X	X	X	X														
New York	X	X	X	X	X	X	X	X	X	X	X	X	X	X	X	X	X	X		
North Carolina	X	X	X	X			X	*	X	X										
Oklahoma		X	X	X	X	X														
Oregon	X	X	X	X	X		X			X										
Pennsylvania	X	X	X	X	X	X*	X			**Xx**										

Rhode Island	X	X	X	X								
South Carolina	X	X	X	X	X		X					
Texas	X	X	X	X	X					**Xx***		
Utah												
Vermont	X	X	X	X	X	X	X			**Xx**		
Virginia		X	X	X						**Xx**		
Washington	X	X	X	X	X	X					X*	
West Virginia	X	X	X	X								
Wisconsin	X	X	X	X	X	X	X	**Xx**	**Xx**	**Xx**	**Xx**	**Xx**

Source: Adapted from Electronics Takeback Coalition. www.electronicstakeback.com.

X: This indicates manufacturers' obligation. It implies that manufacturers must either provide recycling or pay for recycling programs of their electronic waste.
Xx: This indicates that the product is also covered for recycling. It implies that consumers can bring back the products that are marked (**Xx**) for recycling under state programs. This is usually for free.
*

1. In California, only portable DVD players are covered.
2. In Maine, only portable DVD players are covered.
3. In Michigan, fax machines and scanners are covered only if they are part of multifunction.
4. In North Carolina, fax machines covered only if they are part of multifunction.
5. In Pennsylvania, e-readers are covered only if they have a browser and internet capability.
6. In Texas, keyboard mouse is returned with computer and made by the same manufacturer.
7. In Washington, only portable DVD players are covered.

very convenient. A similar result is obtained when a state has a clearly documented collection goal that manufacturers are mandated to meet.

In terms of a well-established collection infrastructure, take the example of states like Oregon and Washington. In these states, we have already seen from Table 2.1 that it is required to have a collection site in every county and city whose population exceeds 10,000 people. According to the Northwest Product Stewardship Council, as many as 92% of residents in Washington has access to a convenient collection site within 10 miles from their place of residence (Northwest Product Stewardship Council).

In terms of collection goals, take the example of Minnesota. According to state law, it is mandatory for manufacturers to meet certain collection goals every year. These goals are correlated with their sales volume in the state during the previous year. During year 1, the goal was 60% by weight and rises to 80% by weight during the second year (www.electronicstakeback.com). Thus, manufacturers who collect less than their goal are required to pay a hefty "penalty" charge in terms of $/lb for every pound that they fall short of the goal. This "penalty" constitutes a higher price than the one they would have to pay by meeting the goal.

2.3.2 Takeaway 2

2.3.2.1 Manufacturers Often Adhere Strictly to State Laws and Will not do any More than Required

Most manufacturers will do only what is required of them by the law in terms of product takeback. This implies that they will strictly adhere to collection goals as spelled out in the state law and will not go over and above what is mandated of them. It has been observed that they often do not make any significant efforts to collect electronic waste if certain targets are not clearly specified in the law. This is even more important given the fact that not all states have spelled out specific collection goals per year. For instance, states like Texas, Missouri, Virginia and Oklahoma have passed laws that require that computer companies operate takeback programs. But these laws are devoid of any specific collection targets. According to data obtained from the state of Texas through the Freedom of Information Act (FOIA), in 2009, the only electronic company that apparently took the law seriously was Del Inc. The company collected 15–18 million pounds across the state. This corresponded to the first year for which data was collected. The second year saw an increase in volumes. Electronic waste collected by a small number of companies such as Dell, Cony, Samsung and Altex (based in San Antonio) accounted for as much as 92% of the volume. On the basis of records obtained for 2010, 78 companies sold computers in the state of Texas. However, only 42 companies collected some e-waste; 36 companies collected absolutely no electronic waste whatsoever. Del collected 10 million pounds (www.electronicstakeback.com). The key takeaway is that state laws should specifically mandate the collection goals and other pertinent measures that electronics manufacturers should undertake in order to run a feasible product takeback program.

2.3.3 Takeaway 3

2.3.3.1 Some States Whose Laws Cover Cost of Collecting e-Waste Have Higher Collection Numbers

E-waste collection numbers in the states of Washington and Oregon are some of the highest in the United States. This is no accident. There is a very specific reason for this. The fact is that there are a variety of collectors in the aforementioned states. These collectors range from private companies such as retailers and recyclers to nonprofits to municipal governments. The state laws for the states of Washington and Oregon also require manufacturers to cover not only the cost of collecting electronic waste but also the cost of recycling it. According to Washington law, collectors are to be fairly compensated by manufacturers for collecting electronic waste. More often than not, local governments are unwilling to participate in collecting electronic waste if they are not compensated, although there are some noteworthy exceptions to this.

Thus, the key takeaway here is that a diverse type of collectors contributes directly to the success of product takeback legislation because doing so covers the cost of collecting end-of-life electronic waste.

2.3.4 Takeaway 4

2.3.4.1 Goals for e-Waste Collection Should be Set High and They Should be in the Form of Minimums

As we have seen in takeaway 3, most manufacturers will do only what is required and nothing more as far as product takeback and recycling is concerned. For example, take the case of Minnesota. It was observed in that state that manufacturers stopped collecting e-waste as soon as they hit their collection goals. This was a problem for many collectors who had overcollected electronic waste. Manufacturers refused to go over the limit as set in the law and therefore collectors were left holding the proverbial "bag." In this case, the bag was a physical entity that contained actual end-of-life e-waste. This was a problem not only for the collectors but also for the consumers. This was to be expected. Collection programs in many cases are free to the consumer only so long as manufacturers are willing to pay for the product takeback. But if the manufacturers were to suddenly refuse to comply, the consumer is stuck with the bill. In the case of takeback programs, the "bill" is in the form of collection fees that must be borne by the consumer. This is a disruptive development since some consumers will refuse to recycle if they were to pay to do so.

In Oregon, for instance, the e-waste collection program was very successful, so successful as a matter of fact that about halfway through the year, it became obvious that companies would easily exceed their collection targets statewide. In response to this, a group of manufacturing companies decided to cut back on their efforts to recycle. They did so by dropping some recyclers from their program. Goodwill was also instructed to cease participation in some collection events (Scott, 2009).

Let us consider what happened in Illinois. During 2010, which constituted the first year of collection, it was required that companies meet a goal of 2.5 pounds of e-waste per person. This goal was supposed to slowly increase over the course of time based on the volume that was collected. However, the goal set during the first year was not mandated. Also, if the actual amount collected turned out to be below the prescribed level, it would be reduced by up to 10%. This was not a very good incentive for companies to cooperate during the first year. In fact, this is exactly what happened. Companies ended up collecting only 2.12 pounds per person (instead of 2.5 pounds per person) although a substantially large variety of products was in fact covered for free recycling.

The key takeaway here is that the goal could be set higher and should be in the form of a minimum instead of a ceiling. This would encourage and incentivize companies to participate and would make the collection program successful. Also, a bill that combines collection goals with convenience requirements could stand a greater chance of success.

2.3.5 Takeaway 5

2.3.5.1 Manufacturers Seem to Prefer Collection in Urban Areas Over Rural Areas

Manufacturers will often prefer to collect electronic waste from urban areas instead of rural areas. One of the reasons for this is that because urban areas have a higher population density, it is easier and more cost effective to collect there. Urban areas will also, in most cases, have a better developed infrastructure to enable e-waste collection. Thus, when the above two reasons are combined, it is clear that it costs the manufacturer less to collect electronic waste in urban areas. For this reason, some states such as Washington, Oregon and New York incorporated some language in their respective bills to mandate collection in every county, not just urban areas. Minnesota, in the other hand, gave manufacturers an opportunity to earn as much as 1.5 times the credit for products that they collected in rural areas to count towards their annual goal.

The key takeaway here is if a state has a large rural expanse, their bills could include language to incentivize manufacturers to collect in rural areas as well.

2.3.6 Takeaway 6

2.3.6.1 The Level of Recycling is Enhanced by Bans on Landfills

When states enact bans on landfills, they are often "rolled out" to coincide with the start of their collection programs. It has been observed that e-waste volume sees a significant spike when the ban actually goes into effect. For instance, consider the example of Maine. The collection program in that state began in January 2006. The landfill ban started only in July of that year. Thus, there were exactly six months for the landfill ban to go into effect since the beginning of the collection program. Collection volumes did not begin to rise until the landfill ban went into effect. The numbers bear it out. The state of Maine collected a total of 1,291,202 lbs of electronic waste during the first six months of 2006. This was in sharp contrast with the following six months when the state collected 2,869,372 lbs of waste.

The key takeaway here is obvious. Collection programs fare much better if they are simultaneously accompanied by concurrent landfill bans which prevent electronic waste from being disposed of into incinerators and landfills.

2.3.7 Takeaway 7

2.3.7.1 Environmentally Responsible Handling of e-Waste is Ensured by Proactive Decision Making on the Part of the Individual States

Over the years, the recycling industry has been plagued by so called rogue companies that try to undercut the state law on e-waste collection and recycling by adopting a host of unsavory measures to manage the collected waste. Such tricks include exporting the waste to developing countries. Others have stockpiled electronic waste in warehouses and then seemingly vanished into thin air without having an ounce of responsibility for the toxic waste they left behind. Yet others just dump the electronic waste in the US. Take the example of a collector who collected computers, loaded them onto a barge and then sunk the barge to the bottom of a lake in the US. Such examples, albeit extreme, still deserve a mention. It is also worth noting that most companies do not adopt such unsavory practices. But whether deliberately or inadvertently, they have adopted work practices that jeopardize the safety of their workers.

In the US, federal laws that regulate the recycling industry seem to be lacking. This void is filled by individual states that have created recycling standards. Companies are expected to strictly adhere to these standards. But, once again, this is easier said than done. The trouble is how do you verify compliance on the part of a company? There are two voluntary standards and certification programs that can help dealing with this issue, namely R2 and e-stewards. We have already discussed these programs in detail earlier in this chapter. Both programs have a similar goal, but there are some subtle differences. For example, while the e-stewards does not permit exporting e-waste to developing countries, the R2 program does allow it. For a variety of reasons, the e-stewards seems to be superior standard. But it is really up to individual states to either pick a particular standard or provide manufacturers with options.

The key takeaway here is that states could either provide manufacturers with an option or mandate adherence to one of the two aforementioned standards. This can be achieved through obtaining certification that demonstrates compliance.

2.3.8 Takeaway 8

2.3.8.1 Remember: Despite the Obvious Advantages of Recycling, Reusing a Product is Always Preferable to Recycling it

As we shall see in Chapter 6, practices such as product reuse and remanufacturing are always preferred over recycling. This is because recycling is always done at the material level. As such it breaks down as much as 90% of the energy embodied in the product. Thus, whenever possible, it is always better to reuse a product.

Many entities in the US ranging from small locally based nonprofit ones to large commercial recyclers reuse and refurbish used equipment so that they can be resold or placed in communities that need such items. Thus, it is important for lawmakers to make certain that laws do not offer a reward (usually in the form of a disproportionate incentive) for encouraging recycling over reusing a product. Take the case of California, for example. In that program, recyclers are reimbursed for the number of units that are recycled, not the ones that are reused. This incentivizes recycling over reusing a product.

On the other hand, consider the case of Illinois. The state incentivizes product reuse over recycling. This is achieved by offering manufacturers an additional credit towards their goal for units that are reused, not recycled. Along similar lines, when equipment is collected through charities whose main role is to reuse a product, the state of Washington offers a bonus. This law used to require companies that did light refurbishing to register as processors. Such a requirement was clearly disadvantageous insofar as product reuse is concerned. Most end-of-life products require some degree of refurbishment (as we will learn in Chapter 6), be it light refurbishment or a substantive one before they can be sold to be reused. However, that legal requirement was later done away with, thus clearing the path for product reuse.

2.3.9 Takeaway 9

2.3.9.1 Consumers Would Like to Bring Back "all" Electronic Products not Just a Few Selected Ones

Most state laws specify the types of products also referred to as the "scope of products" that can be returned for free recycling. We have already learnt about this in Tables 2.1 and 2.2. Initially some of the laws that dealt with tackling the problem of electronic waste often specified that only a few products such as computers, laptops and monitors could be returned. This was mostly due to political limitations at the time. However, more recently passed laws have been able to specify that a variety of products can be brought back for free recycling. Evidence suggests that consumers would like to bring back a much larger variety of electronic items such as big-screen TVs and printers. In states such as Washington and Oregon, TVs represent over 60% of the e-waste being brought back by consumers.

The key takeaway here is it would be preferable to expand the list of electronic items that can be recycled for free in order to make the product takeback program a success. This is also what consumers seem to prefer.

2.3.10 Takeaway 10

2.3.10.1 Product Takeback Programs Should be Transparent

Most companies across the US do not voluntarily report the volume of electronic waste they collect in each state. Despite their promises to run a vigorous product takeback program, there is really no way to ascertain the degree of success companies are able to achieve. This information is certainly not available to the public. Consider the example of the state of Texas. Although companies do report their takeback volume to the state, Texas law does not require that this information be made public.

In order to obtain such information, it is required for an NGO to file a Freedom of Information Act (FOIA) request every year. According to numbers obtained for 2009 and 2010, companies were making widely varying levels of efforts to collect e-waste. In order to rectify this, each state could publish a report on the amount of volume each manufacturer collected every quarter or so. This would impart to the product takeback program a higher degree of transparency. It would also allow local residents an opportunity to make comments and offer suggestions.

2.4 CONCLUSION

In this chapter, important environmental legislations from several US states and different countries globally were presented. These legislations were compared with each other and their scope of coverage was determined. This chapter forms an important basis to answer the question: "Why should we even bother to design and develop environmentally sustainable products?" Limited and valuable natural resources in combination with ever stringent government legislation should be able to sufficiently answer this question. We will address product design goals in the next chapter.

REFERENCES

Caterpillar Reman: www.caterpillar.com/en/company/sustainability/remanufacturing.html

Choi Athelstan, K.Y. (1994), "A screening method for life-cycle inventory analysis for industrial material," Master of Applied Science Thesis, University of Windsor, Ontario, Canada.

Covered Electronic Device Act, State of Illinois: www.ilga.gov/legislation/publicacts/95/PDF/095-0959.pdf

Daimler Recycling: https://www.daimler.com/sustainability/resources/from-old-to-new.html

Dell Takeback Program (Bloomberg NEF): https://about.bnef.com/blog/dell-eyes-63-billion-e-waste-recycling-opportunity-qa/

Electronics Takeback Coalition: www.electronicstakeback.com

Environmental Law in China, BBC: https://www.bbc.com/news/world-asia-china-48847062

E-Stewards Standard: www.estewards.org

European Union Environmental Legislation: (https://ec.europa.eu/commission/presscorner/detail/en/MEMO_05_248

Feldmann, K., Traunter, S. and Meedt, O. (1999), "Innovative disassembly strategies based on flexible partial destructive tools," *Annual Reviews in Control*, Vol: 23, pp: 159–164.

Ford Core Recovery Program: https://www.fenderbender.com/articles/3513-core-charge-programs-ramping-up

Illinois Environmental Legislation: www.ilga.gov/legislation/publicacts/95/PDF/095-0959.pdf

McKenna, R., Reith, S., Cail, S., Kessler, A. and Fichtner, W. (2013). 'Energy savings through direct secondary reuse: An exemplary analysis of the German automotive sector," *Journal of Cleaner Production*, Vol: 52, No: 1, pp: 103–112.

Nagel, C. and Meyer, P. (1999), "Caught between ecology and economy: end-of-life aspects of environmentally conscious manufacturing," *Computers and Industrial Engineering*, Vol: 36, pp: 781–792.

New York Environmental Legislation: www.dec.ny.gov/docs/materials_minerals_pdf/ewaste law2.pdf

R2 Directive: sustainableelectronics.org

Ron Ad de and Penev, K. (1995), “Disassembly and recycling of electronic consumer products: An overview,” *Technovation*, Vol: 15, No: 6, pp: 363–374.

Scott, L. (2009), “Oregon’s electronics recycling too successful for some manufacturers.” *The Oregonian*. March 12, 2009.

Virginia Environmental Legislation: https://law.lis.virginia.gov/vacode/10.1-1425.31/ 6a. https://law.lis.virginia.gov/vacode/10.1-1425.30/

6b. https://law.lis.virginia.gov/vacode/10.1-1425.29/

WEEE Directive: https://eur-lex.europa.eu/legal-content/EN/TXT/?uri=CELEX:32012L0019)

3 Design for Disassembly

3.1 IMPORTANCE AND DEFINITION OF DESIGN FOR DISASSEMBLY

According to Gungor and Gupta, disassembly is defined as "a systematic process that allows reusable, non-recyclable and hazardous subassemblies to be selectively separated from recyclable ones" (1997). Such a process strives to recycle material efficiently. In order to achieve this, proper separation of materials is necessary. Different disassembly strategies are evaluated and the best among them is eventually chosen. Another definition provided by Brennan et al. is "the systematic removal of desirable constituent parts from an assembly while ensuring that there is no impairment of parts due to the process" (1994). Brennan considers nondestructive disassembly since it is known to have higher recycling value. According to Kriwet, disassembly is the decomposition of an assembly into subassemblies or components while keeping their original form (1995). In the engineering context, products that have reached the end of their useful life are taken apart. This is done primarily in order to recover useful components by regaining the value added to products and materials (Ron, 1995).

In the present era of environmental awareness, end-of-life (EoL) objectives such as component reuse (components from a retired product being used without upgradation in a new product), remanufacture (components from a retired product being used in a new product after technological upgradation) and recycling (reuse at the material level: e.g., recycling of plastics) constitute some of the more important reasons for disassembling products. The concept of reuse, recycling and remanufacturing a product shall be dealt with in detail in chapter 6 of this book.

To put things in perspective, the volume of electronic waste in Germany exceeded 800,000 tons annually in the early 1990s. Since the 1990s, this problem has only become worse, primarily on account of rapid economic development of the BRIC (Brazil, Russia, India and China) nations. Around 600,000 refrigerators are discarded annually in the Netherlands while about 1.5 million automobiles are transported to dismantling plants every year in the United Kingdom. The stock of potentially unrecoverable materials embedded in common domestic appliances such as washing machines and refrigerators is extremely large especially when these products/components have not been designed specifically with recycling/reuse in mind (Lee et al., 2001). For example, it has been estimated that around 33 kilotons of electronic products were discarded in the Netherlands in 1992 alone. This scrap included 8 kilotons of TV sets, 8.5 kilotons of audio apparatus, 3 kilotons of computers and 1.5 kilotons of telephone equipment (Ron, 1995). Industrial waste is only set to grow exponentially (industrial waste will increase to about 57 kilotons by the year 2005) at this rate.

The problem pertaining to electronic products is especially severe. With the volume of electronics waste rising rapidly, there has been increasing concern over the number of electronics items ending up in landfills. These items contain toxins such as lead, mercury, cadmium, chromium and so on. The situation has deteriorated to the extent that at least two dozen US states have now passed laws concerning e-waste.

For instance, the Covered Device Recycling Act of 2010 passed by the state of Pennsylvania initiated the landfill ban in January 2013. Under this law, manufacturers are responsible for proper disposal of electronics. It is imperative that manufacturers register with the state to sell their products as well as establish and conduct ongoing recycling programs (CDRA 108, 2010).

We have cited just two examples here that underline the gravity of the situation created by waste. Most of this waste is generated due to the extant widespread consumerism in the world. Reduction, recycling and remanufacturing is the only sustainable solution to managing this waste. The topic of product recycling and remanufacturing has been dealt with in Chapter 6 of this book. The more conventional approach to dealing with consumer product waste is to dispose it in landfills.

However, the number of potential landfill sites for nonhazardous solid wastes has seen an exponential decrease. In the United States alone, landfill sites have diminished from 18,000 in 1985 to 9,000 in 1989. According to a recent study, the United states had lost more than 70% of its landfill sites by 1997 (Zhang et al., 1997) with landfills in many states reaching their permitted capacities at an alarming rate. End-of-life products contain extensive amounts of reusable material, which is too expensive to dispose of, but retrieval of which would be beneficial to the manufacturer as well as the environment.

Proponents of sustainable design and green manufacturing would be delighted to learn that a prototype car conceived entirely from vegetable fibers has been designed jointly by Benz and Swatch. German automaker BMW has been following a different approach towards achieving the same goal. Color-coding for different plastic materials has been used actively by BMW during the past 15 years. The scheme facilitates development of efficient disassembly and dismantling techniques.

Knowledge thus gained is transferred back to new product design and development. BMW is essentially looking at a car manufactured out of 100% recycled parts in the near future (Hundal, 1996). Swedish auto manufacturer Volvo (Benson, 1996) is relying more with its contact with universities and academia. It intends to improve the efficiency of its recycling method awarding research grants to specific projects including life cycle analysis, material recycling, disposal of environmentally hazardous materials, dismantling methods, energy recovery and transportation of materials to and from recycling centers.

In view of the obvious importance of the disassembly process for sustainable product design and manufacturing, it is imperative to evaluate this process in depth. We have already defined it in the previous paragraph. The process of disassembly may be classified into two broad categories based on the method of disassembly:

- Destructive disassembly or brute force approach.
- Nondestructive disassembly or reverse assembly

Depending on the extent of disassembly, nondestructive disassembly can be further classified into two categories as follows:

- **Total disassembly**: The entire product is disassembled into its constituent components. This is sometimes not economically feasible due to the imposition of external constraints such as time, economic factors, presence of hazardous materials and so on.
- **Selective disassembly**: Products are deemed to be selectively disassembled when they are subjected to reversible dismantling. This process leads to the formation of subassemblies that are comparatively less complex. It could also result in the obtainment of single parts (Lambert, 1999). In the process of selective disassembly, desirable constituent parts are systematically removed from an assemblage of components (assembly), while simultaneously making sure that no parts are impaired or damaged either during the process or due to it (Brennan et al., 1994). The main goals of the process of selective disassembly are enumerated as follows:
  1. Products are often selectively disassembled in order to enhance the ease with which maintenance and repair activities (serviceability) can be accomplished. All service-related operations such as maintenance, diagnosis and repair and malfunction are included in this list. Products that are designed for ease of serviceability result in several benefits such as reduction in warranty costs, which leads to higher EBIT, greater degree of customer appeal and longer service life. Service life is extended on account of the product's ability to be serviced while incurring lower expense.
  2. Subassemblies are available either in the form of service parts or they can act as components that can be assembled into new products. Economies of scale can be achieved by reusing components that are still robust enough for extended use in another product. This is referred to as cost optimization. Such components are reused in new products instead of being recycled.
  3. Parts or components are removed before other desired parts are set free.
  4. Availability of parts intended for material reuse (recycling) is ensured.
  5. Removal of contaminants results in increased purity of materials.
  6. Selectively disassembling products facilitates achieving compliance with environmental regulations. These regulations prescribe removal of particular parts, materials and substances for reasons of environmental conservation and safety. For instance, removing working fluids such as engine oils, lubricants and so on are an example of this goal.

3.2 DISASSEMBLY PROCESS PLANNING

It is imperative to perform comprehensive disassembly process planning in the early stages of product design. This planning should be done in order to render a product easy to disassemble. The overall concept of a disassembly process plan is illustrated in Figure 3.1.

Any disassembly process plan begins with a given product group that needs to be disassembled. Certain characteristics are found to be common between every product group and can be aggregated. Four essential stages need to be incorporated into every disassembly sequence plan (DSP). These stages are described as follows:

1 **Analysis of the product:** Product analysis is the first step of a disassembly sequence plan. During this step, the end value to be realized by disassembling the product is assessed. This value consists of consideration of the potential of the product to be reused, the amount of value that can be recovered, risk potentials and the state of recycling technology as it currently exists. This last consideration is a very important factor since every subsequent recycling decision depends on this factor.

2 **Analysis of the assembly process:** The process of product disassembly is a function of the assembly process. Thus, in order to truly understand the process of product disassembly, it is first important to understand how the product has been put together. Analysis of product assembly takes into account the types of fits, adhesives, bonding members, grooves, slots and so on that are responsible for keeping various constituents of the product structure together. Knowledge of these parameters makes it easy to plan for disassembling the product. Process parameters include determination of tooling requirements, amount of force, time and manpower necessary to effect disassembly. Knowledge of what components are functionally valuable is also an important aspect of disassembly process planning. Assembly analysis consists of analyzing joining elements, hierarchy of components and sequence of component assembly.

3 **Analysis of usage, mode and effect:** Most commercially available products are disassembled at the end of their useful lives. As such, they have already been used extensively and have been subjected to considerable wear and tear. Sometimes, unintended modifications may have been introduced into a product. This is often the result of the manner in which a product has been used. Such modifications directly affect the intensive and extensive properties of the product, which often have a detrimental impact on the product since they were not intentionally incorporated into the design by the designer. It is important to have knowledge of such modifications in order to plan for contingencies as they arise during the disassembly process.

4 **Strategic analysis of the dismantling process:** How a product is to be disassembled is decided largely by the reason for disassembling that product. For instance, if the goal of product disassembly is recycling, (at the material level), disassembly need not be carefully performed. In this case since only

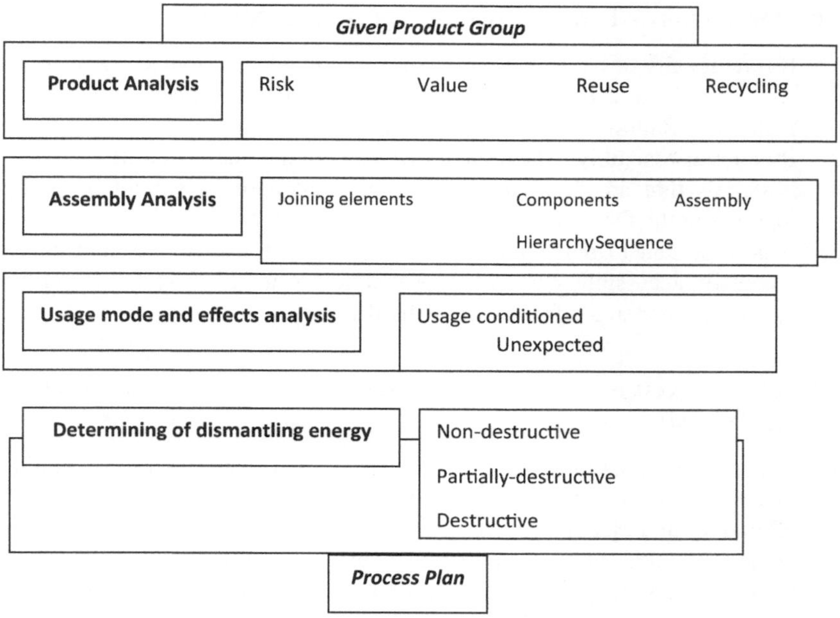

FIGURE 3.1 Disassembly process plan.

the intensive properties of components are important, destructive disassembly may be used to expedite the process. On the contrary, if certain components of the product are to be reused (which is often the case with functionally important and valuable components), care has to be taken in separating that component from the product structure in order to maintain both its intensive as well as extensive characteristics. In this case, nondestructive disassembly may be used to significant advantage.

3.3 DESIGN FOR DISASSEMBLY GUIDELINES

While there are a large number of practically applicable guidelines available on the topic of design for assembly, the same cannot be said for design for disassembly. There are some guidelines available on this topic, but they have not been complied together into what can be regarded as a systematically applicable set of tools. Most of these guidelines aim to achieve three main objectives:

- Minimize product variety.
- Utilize modules in product design and construction.
- Similar materials should be clustered together.

Some of the more important disassembly guidelines are enlisted in this section.

3.3.1 The Need for Extensive Assembly Should be Minimized

1 If functionally possible, similar elements should be combined together.
2 Disassembly procedures can be predicted with a greater degree of certainty if material variability is minimized as much as possible.
3 Product disassembly is facilitated by using compatible materials.
4 Materials that are functionally important but harmful should be grouped together in the form of subassemblies so they can be disposed of quickly.
5 It is important to make sure that any valuable, reusable and harmful parts are easily accessible. Incorporating this design feature results in significant savings in time and effort in accessing the part.

3.3.2 Configuration of the Product Should Always be as Predictable as Possible

1 Material combinations that lead to rapid ageing and corrosion should be avoided. This is because it is often difficult to disassemble such materials efficiently and cleanly. Materials that are subject to corrosion often have a tendency to not only corrode but also spread corrosion and break off inside the product. Avoiding such materials results in a product configuration that is more predictable as far as disassembly is concerned.
2 Subassemblies can be protected from corrosion by following the same design guidelines as explained earlier.

3.3.3 The Process of Disassembly Should be Facilitated as Much as Possible

1 Incorporating easily accessible drainage points in the product design facilitates easy disassembly.
2 The choice of fasteners should be such that they are easy to remove (selective disassembly) or destroy (destructive disassembly).
3 Removing and unfastening fasteners is a time-consuming process. Disassembly time and effort is minimized by reducing the number of such fasteners as much as possible. Easy access needs to be provided to disjoining, fracture and cutting points.
4 When designing a product for ease of disassembly, special attention needs to be paid to ascertaining the disassembly path. The more convoluted this path is, the more difficult the disassembly process. As a rule, the disassembly path should be simple and straightforward. It should comprise a simple sequence of operations which will result in the removal of most components. The process of disassembly is simplified by avoiding multiple directions and complex movements.
5 Metal inserts should always be avoided in plastic parts since it increases material variety, part complexity and necessitates multiple directions and complex movements to affect disassembly.

3.3.4 Product Should be Easy to Handle

1 Products and components cannot be handled if there are no surfaces that can be easily grasped. Thus, handling is simplified by designing the products such that at least one surface is available for easy grasping.
2 Wherever functionally possible, it is important to avoid nonrigid parts. These parts can move, twist and bend and thereby hinder the disassembly process.
3 Poisonous substances are notorious for causing health hazards. Examples of such substances include lead, mercury and hexavalent chromium, among others. They should be avoided if at all feasible. If not, they should be provided in sealed units to minimize their detrimental impact on human health.

3.3.5 Components Should be Designed so They Are Easy to Separate

1 Processes such as painting and coating constitute secondary processes and often interfere with critical disassembly processes such as accessing components and removing them. For this reason, such processes should be avoided.
2 Disassembling different materials often leads to confusion during the disassembly process. For this reason, such materials should be marked.
3 During selective disassembly, care should be taken to not damage any equipment or machinery. This goal is facilitated by avoiding parts and materials that are likely to cause such damage.

3.3.6 Part Variability Should be Reduced

1 Reduction in parts variability is achieved by using standard subassemblies and parts. This practice facilitates modular construction of products.
2 If possible, product design should incorporate the least variety of fasteners possible. Examples of fasteners include screws, nuts, bolts rivets and so on. The larger the variety of fasteners used in a product, the more the need for an equally greater variety of tools to disassemble them, skills to access and remove the fastener, surface preparation for disassembly and working postures necessary for achieving the disassembly process.

3.4 PRODUCT MODULARIZATION FOR DISASSEMBLY: DESIGN APPROACH TO DISASSEMBLY

Nondestructive disassembly of products is performed with the objective of achieving ease of recycling and remanufacturing. Doing so leads to an increase in the recovery rate of products by allowing selective separation of parts and materials (Gu, 1999). The proactive approach to solve the disassembly problem is to consider disassembly at the product design stage itself. Life cycle analysis indicates that a large chunk of the entire cost associated with the product can be attributed to the product design process.

Design for disassembly (DfD) initiatives lead to the correct identification of design specifications to minimize the complexity of the structure of the product (Gungor and Gupta, 1999) by achieving the following objectives:

- Minimizing the number of different parts.
- Increasing the use of common materials.
- Optimal use of removable yet robust fastener and joint types. An ideal design may maximize the use of a single robust joint by having many parts come loose when that single joint is separated (Kuo, 2000).
- Optimizing the spatial alignment between various components to facilitate disassembly without jeopardizing assemblability, functionality and structural soundness.

In the long run, this could make it possible to eliminate the need for land filling and incineration of mixed waste. Products would be designed so as never to become waste, but instead act as inputs into new products at the end of their useful lives. Design for disassembly is a key strategy within the larger area of sustainable product design: it is concerned with a more proactive approach to environmentally responsible design (Krikke et al., 1998).

Disassemblability is defined as the degree of easy disassembly (Mok et al., 1997). Determining factors for disassemblability have been outlined based on the disassembly mechanism. A case study for the (re)design of an automotive instrument panel using the design and quantification methodology is also presented in this chapter. The following factors affect disassemblability:

- Use of force. Minimal use of force is recommended.
- Mechanism of disassembly. A simple mechanism is preferable.
- Use of tools. Ideally, disassembly should take place without the use of tools. If needed, the task should be doable using simple tools. Complex tools are generally not preferred.
- Repetition of parts. As far as possible, component parts of a product should not be repeated. This will aid in quick and easy identification of parts at each stage of disassembly thereby reducing confusion.
- Recognizability of disassembly points. Disassembly points are defined as those joints which need to be disjointed to affect disassembly. Easy recognizability of such points is advisable especially in the case of complex product structures or products that incorporate snap fits. Similarly, easy recognizability of disassembly points assumes a high degree of importance in the case of products that accumulate internal dirt during their useful life.
- Product structure. The simpler a product structure, the better it is from the disassembly point of view.
- Use of toxic materials. It is advisable not to incorporate toxic materials in the design of parts since they may pose health hazards to the operator performing the disassembly. This is especially important since most disassembly is still manual in nature.

When a product is designed from the design for disassembly perspective, several changes may be incorporated in the product which would render it technically faulty or structurally unsound. Product redesign would be necessary in such a situation. This may be affected by redesigning components, standardizing parts, materials and subassemblies, devising innovative joining methods and so on.

Gu and Sosale (1999) developed a methodology to modularize a product for life cycle analysis (LCA). The proposed methodology considers various Life Cycle Engineering (LCE) objectives such as assembly, maintenance, reuse and recycling. It is also important to note that maintenance, reuse and recycling involve varying amounts of disassembly in order to be performed successfully. Modules may be formed considering two approaches:

- First, to form modules based on considering each objective separately and then make tradeoff decisions between various modular configurations.
- Second, modularize a product based upon the weighted average objective considering all modularization objectives simultaneously.

Modularization may have one or more objectives as already discussed. The proposed methodology consists of three main phases:

- **Problem definition**: The problem definition phase consists of steps such as identification of design problem characteristics. Examples of this may be the degree of design originality, product variety, life span of different modules and the scale of the design project. Gu and Sosale (1999) reiterated the fact that decomposition of a product design into functional and physical structures is a prerequisite for product modularization.
- **Interaction analysis**: Parts are clustered (module formation) such that interactions between modules are minimized. The next step is the identification of relevant interaction factors for each modularization objective. For instance, life expectancy of components may be considered as a main factor for reuse. In addition, each interaction factor is decomposed into subfactors. For example, the physical factor may be further subdivided into contact, alignment and containment. To evaluate the interactions for the objective, values are assigned for each interaction. A hierarchy of objectives, factors and subfactors is thus formed. This leads to the formation of an interaction matrix containing the values of all interactions among all components. An interaction value indicates the degree of importance that two components have between them. These values are then evaluated on a scale of 1–10. All interaction values at the lowest level are used collectively to calculate the weighted sum of interactions for an objective. All weights representing the relative importance of factors and objectives are normalized such that summation of all branches from each node equals one.
- **Module formation**: Finally, module formation is done to maximize the interactions between components within modules. The clustering algorithm randomly generates several configurations in search of the optimal configuration. As a new random configuration is generated, objective function values

are recalculated for new configurations. The new configuration replaces the previous one in the memory based upon probability of change. Gu and Sosale (1999) also presented two case studies: one for a vacuum cleaner and another for a starter in support of their algorithm.

Shu (1999) presented an application of a design-for-remanufacture framework to the selection of product life cycle fastening and joining methods. Remanufacture is a production batch process constituting disassembly, cleaning and refurbishment or replacement of parts of products at the end of life (Lund, 1983). As such remanufacturing involves recycling at the parts level as opposed to the scrap materials level. Recycling at the higher level of components avoids resource consumption for possible unnecessary reprocessing of materials while preserving the value added during manufacturing. The following disassembly guidelines are presented to aid the designer:

- The quantity and variety of fasteners should be reduced.
- Standard fasteners should be the preferred material.
- It is advisable to avoid long disassembly paths and awkward access.
- If dirt is expected to be a problem at the time of disassembly, fasteners should be selected such that even dirty ones are easy to locate and remove.
- Many design-for-assembly guidelines such as those facilitating part handling and access are beneficial even from the disassembly perspective.
- It is often easier to take a set of components from the higher state of order in the assembled state to a lower state of order in the disassembled state than doing the reverse.

The issue of accessibility of joints not otherwise used in routine maintenance operations is addressed at length. In this context, the ever-popular snap fits are examined critically. Shu argued that snap fits are especially useful in situations where only the recovery of material is important. In situations which entail the recovery of a component intact, the snap fit may not be very useful in that it might damage the part in question and render it useless. Situations such as the one described above are especially common when a product is being taken apart to recover its components for remanufacture.

The overall concept of modular products is described by Ulrich (1995) by proposing a methodology to modularize components by grouping function-related elements together. Ulrich uses the concept of product architecture: A scheme by which the function of a product is allocated to physical components. Modular architectures may be divided into three distinct subtypes: Slot, bus and sectional. A brief synopsis of each type of architecture follows.

- Slot: Each of the interfaces between components in slot type architecture is of a different type than the others thereby preventing interchangeability.
- Bus: In this type of architecture, there is a common bus to which other physical components connect via the same type of interface: For example, an expansion card for a personal computer. Automobile chassis is another example.

TABLE 3.1
Objectives and goals of design for disassembly

Design for disassembly	
Facilitate easy access to components	Optimization of disassembly sequence
	Design parts for easy removal
	Avoidance of embedded parts
Simplify component interfaces	Avoid springs, pulleys and harnesses
	Avoid adhesives and welds in design
	Avoid use of threaded fasteners
Design for simplicity	Reduce complexity of products
	Reduce number of parts
	Design multifunctional parts
	Use common parts where possible

Source: Adapted from Zhang et al. (1997).

- Sectional: In this type of architecture, all interfaces are of the same type and there is no single element to which all other components attach. The assembly is built up by connecting the components to each other through identical interfaces: For example, sectional sofas and office partitions.

Zhang et al. (1997) provided guidelines to enable the designer to design for disassembly. These guidelines are presented in Table 3.1.

3.5 DISASSEMBLY ALGORITHMS AND SEQUENCE PLANS: REACTIVE APPROACH TO DISASSEMBLY

A disassembly sequence plan is a sequence of disassembly tasks that begins with a product to be disassembled and terminates in a state where all the desired parts of the product are disconnected (Gungor and Gupta, 1998). A disassembly sequence plan to optimize product recovery through the minimization of cost, maximization of material recovered, minimization of disassembly time using mathematical techniques such as linear programming, dynamic programming and graphical tools.

Thierry et al. (1995) proposed a product recovery management (PRM) approach wherein returned products can be recovered at four levels: product, module, part and material level (in that order). Product recovery options achievable by disassembly may be classified into categories as enumerated in Table 3.2.

The objective of this methodology was to recover as much as possible the economical as well as ecological value of products, components and materials so as to reduce the ultimate quantities of waste to a minimum. One of the drawbacks of this approach is that the distinction between refurbishing and remanufacturing is too subtle to be implemented from the design perspective.

TABLE 3.2
Product recovery options after disassembly

Options	Objective	Level of disassembly	Result
Repair	Restore working condition	Product level (Limited disassembly and fixing)	Some parts repaired
Refurbishing	Improve quality level (though not like new)	Module level (Some technological upgrading)	Some modules repaired / replaced
Remanufacturing	Restore quality level as new	Part level	Used and new parts in new products
Cannibalization	Limited recovery	Selective disassembly and inspection of potentially reusable parts	Parts reused and /or recycled/ disposed of
Recycling	Reuse materials only	Material level	Materials used in new products

Source: Thierry et al. (1995).

A business case study at Roteb, the municipal waste company of the city of Rotterdam, the Netherlands, is presented by Krikke et al. (1999). The study concerns the determination of an optimal recovery strategy of discarded computer monitors. The aim was to analyze the economic viability of monitor recycling on the one hand and to validate the practical viability of the models developed in earlier research on the other hand. The first step in determining the optimal recovery strategy consists in initially forming a disassembly tree. The second step consists in assigning optimal recovery and disposal options to the product (in case of no disassembly) or the released components (in case disassembly of components is done). Recycling options are classified into two distinct categories: mixed option and separate option. The mixed option is applicable to the product as a whole whereas the separate option is applicable to the individual components. They are usually high-quality recycling options. These options are specified as per material clusters and each part in this cluster is processed accordingly. Roteb implemented a feasibility criterion for the application of reuse, recycling and disposal, an expansion of which is depicted in table. The following return and disposal (RD) options were available to process a return monitor.

- Product: Disassembly or mixing process.
- Modular unit: Disassembly or mixing process.
- Parts: Separate processing per material cluster.

The problem of dealing with future uncertainties of recycling options is addressed by Zussman et al. (1994) using the utility theory. Attributes such as technology refinement, prices or dumping fees are bound to change with time (dynamic) and as such involve uncertainty for the designer/policymaker. The methodology incorporates the optimization of disassembly sequences to achieve objectives such as maximization of net profit, maximization of parts reuse or minimization of waste headed towards landfills. A recovery graph is constructed using the AND/OR logic as described earlier. Values of different entities such as disassembly sequences, recycling options and related costs are used to plot the recovery graph. The principal drawback of this approach is it does not specify the time horizon over which future uncertainties are being planned. Apparently, recycling options are bound to change little over a shorter period of time. In the case of a longer timing horizon, product design itself would be subject to numerous changes due to technological refinement. These and related issues have only been vaguely addressed.

Two algorithms were presented by Gupta (1997) to obtain a scheme for disassembling multiple product structures having common parts. The core algorithm determines the number of root items to be disassembled in order to minimize disassembly cost. A root is defined as a major subassembly which is further composed of minor subassemblies that can be subdivided into individual components. The allocation algorithm is used to determine a disassembled schedule for the roots and subassemblies by allocating the disassembly requirements over the planning horizon.

Gungor and Gupta (1997) presented an evaluation methodology to choose the best disassembly process from among several alternative processes. Disassembly sequences are generated heuristically by considering the following factors: Precedence relationship of components of the product under consideration and the average difficulty ratings for each component of the product. Total time of disassembly (TTD), a metric affected by attributes such as joint type and direction of disassembly changes, was chosen for evaluating a particular disassembly sequence.

Approaches dealing with disassembly sequence planning present only one facet of the possible solution. For example, the approach described by Navin Chandra concentrates solely on earning revenue from disassembly. It does not consider such possibilities as an optimal disassembly sequence being time consuming, hazardous to the workers performing the disassembly and so on. As such, nonfinancial yet practically important factors have not been considered and the approach remains merely theoretical. The methodology described by Kuo failed to clearly define "a valuable component." A component may be valuable in terms of its monetary value, utility value or recycling value. The methodology suggested by Penev and Ron doesn't seem to recognize other evaluation metrics such as effort required for disassembly, material compatibility and so on which play an equal if not a more important role in design evaluation.

A high degree of uncertainty is introduced into the disassembly process due to upgrading/downgrading a product during its use by customers and due to defects occurring in use or during disassembly. Gungor and Gupta (1998) have proposed a methodology for the disassembly sequence planning of products with defective parts. Changes in disassembly sequence plans have to be incorporated in order to handle factors leading to uncertainties.

Availability of original CAD drawings and an unchanged product structure has been assumed. The physical relationships among components is represented using the disassembly precedence (DP) matrix developed from the original CAD model of the product using the AND, OR and OR/AND relationships. Next, an optimum disassembly sequence plan is generated. Finally, the actual disassembly process is

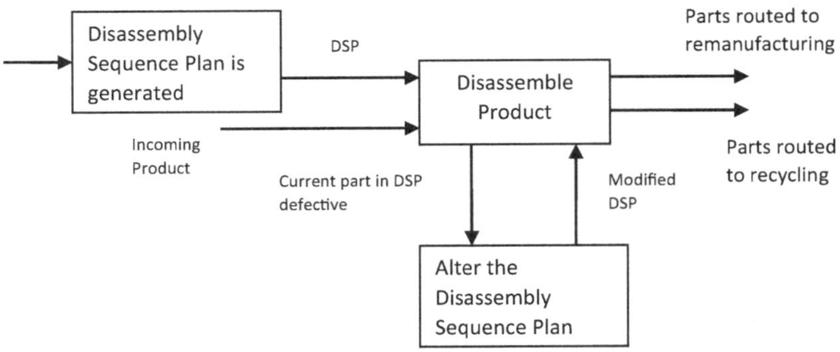

FIGURE 3.2 Uncertainty handling in disassembly sequence planning (Adapted from Gungor and Gupta, 1998).

performed. An unexpected situation is dealt with by appropriate modification of the disassembly sequence plan.

The issue of disassembly costs has been touched upon by Feldmann et al. (1999). However, from the recycling point of view, recycling costs and benefits differ for specific fractions of recovered materials. Disassembly costs must be justified by the economic advantages of recycling. The more important economic considerations to be considered during the disassembly process include the following (Ron, 1995):

- The value added to products and materials during manufacturing.
- The disassembly cost per operation.
- The revenue per operation.
- The penalty if poisonous materials are not completely removed.

In order to determine the optimal disassembly and recycling strategy, all available disassembly combinations have to be compared. To this end, operating costs continue to be one of the most daunting concerns for manufacturers. The end-of-life economic value of components can be computed using the following methods (Lee et al., 2001):

- Miscellaneous cost: Collection cost + processing cost.
- Reuse value: Cost of component – miscellaneous cost.
- Remanufacture value: Cost of component – remanufacture cost miscellaneous cost.
- Primary recycle value: (Weight of component × market value of material) – miscellaneous cost.
- Secondary recycle value: (Weight of component × scrap value of material) – miscellaneous cost.
- Incinerate value: (Energy produced × unit cost of energy) – miscellaneous cost.
- Landfill cost: (Weight of component × cost of landfill) – miscellaneous cost.
- Special handling cost: (Weight of component × cost of special handling) – miscellaneous cost.

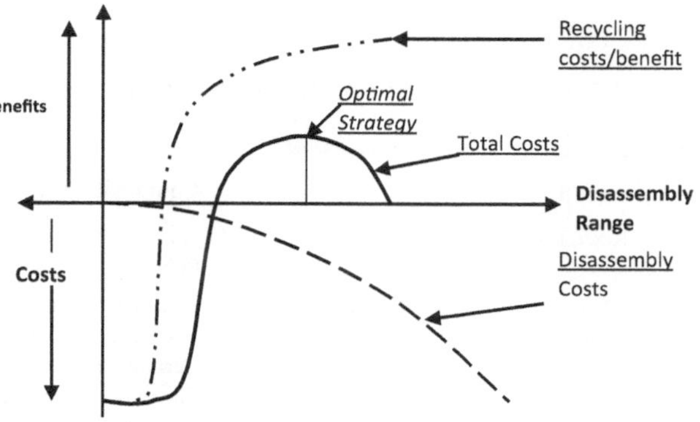

FIGURE 3.3 Determination of optimal recycling and disassembly strategy (Adapted from Feldman et al. 1999).

A graphical illustration depicting the role played by various economic factors in determining an optimal disassembly and recycling strategy is presented in Figure 3.3.

Ron and Penev (1995) devised a disassembly strategy based on the economic validity of the disassembly process. They suggested the following steps:

- Determination of the value of the abandoned product. This value has been represented as: $V_{ab} = (R_{dis} + R_{shr} - C_{tr} - C_m)$. The following terminologies have been defined:
  V_{ab}: Value of the abandoned product.
  V_{pr}: Value of the product
  R_{dis}: Revenue from the disassembled parts.
  R_{shr}: Revenue from recycling the remaining materials.
  R_{dm}: Revenue from dismantling. Optimal strategy total costs disassembly costs disassembly range recycling costs/benefits costs benefits.
  C_{tr}: Transportation costs.
  C_m: Miscellaneous costs (testing, remanufacturing and quality assurance costs).
  C_{dis}: Disassembly cost to repair the product.
  C_{as}: Assembly cost to repair the product.
  C_c: Value of the component replaced.

  State of the product with respect to its life cycle at the moment it is discarded: If a product is not completely out of its operational life cycle, it can be brought back into operation with minimum effort by appropriate servicing. This is recommended if the product is in the second phase of its life cycle. Servicing should be done if the following equation holds true:

$$V_{pr} - (C_{dis} + C_{as} + C_c + C_m) > (R_{dm} + R_{shr} + R_{dis}).$$

Otherwise, it was recommended to proceed to the next disassembly level.

TABLE 3.3
Type of recycling according to component composition

Adaptation of recycling to composition of component

Type of recycling	Definition	Composition of the component
Primary recycling	Recycling on a comparable quality level	No alloy present in the component Polymer content in the component
Secondary recycling	Recycling on a lower quality level *Down cycling*	Presence of an alloy in the component No polymer content Ceramic content Elastomer or composite material
Tertiary recycling	Decomposition	
Quaternary recycling	Incineration with energy retrieval	No polymer content Ceramic content

Source: Adapted from Lee et al. (2001).

Lee et al. (2001) devised a multiobjective methodology for evaluating product end-of life options and disassembly. Guidelines for determining feasible end-of-life options including the economic value of products and their components have been discussed. Two end-of-life disassembly charts emphasize the practical application of the proposed methodology. Whereas one of the charts can be used as a tool to assist in product design, the other can be used to determine the optimal stage of end-of-life disassembly of the product. Different guidelines have been suggested to determine feasible end-of-life (various types of recycling, remanufacturing, reuse, incineration, etc.) options for components. They have been presented in Table 3.3.

The following objectives were evaluated using the economic guidelines given earlier:

- To minimize environmental impact by extraction of reusable, recyclable and toxic components from the product. This process is capital and time intensive.
- To minimize deficit or maximize surplus by discontinuing disassembly when the greatest positive net cost (or lowest negative net cost) is attained.
- To minimize the time for disassembly and deficit by discontinuing disassembly when the highest rate of return is achieved.
- To minimize the time for disassembly and environmental impact by discontinuing disassembly when the highest rate of environmental impact is achieved.
- To minimize the environmental impact and deficit by discontinuing disassembly when the highest rate of return is achieved.
- To minimize the time for disassembly, environmental impact and deficit by discontinuing disassembly when the highest rate of return is achieved

The issue of disassembling a geometrically constrained assembly of components was addressed by Srinivasan and Gadh (2002). Two types of constraints were considered: Spatial constraints due to three-dimensional geometric interactions between components and user-defined constraints imposed by clustering components into subassemblies. The problem was solved utilizing an algorithm that focuses on removing components simultaneously instead of sequentially (global selective disassembly algorithm: GSD).

Another CAD-based selective disassembly algorithm was presented by Srinivasan et al. (1999). This algorithm is actually an extension of the one described earlier. After the disassembly sequences have been determined using the wave propagation algorithm, they are evaluated using an objective function such as say minimization of cost. Disassembly design decisions are then based upon this evaluated disassembly sequence.

The disassembly process is still largely manual in nature. However, several researchers are exploring avenues for automating disassembly systems in the search for better efficiency and cost effectiveness. The issue of automation of disassembly and recycling technology has been dealt with in a recent Delphi study. Obsolescence of end-of-life technology motivated this study. Since electronic products and automobiles have an average lifespan of 15–25 years, end-of-life technologies would have changed by the time these products are finally discarded. The need to gain an insight into the nature of future recycling and disassembly technologies is therefore crucial. The conclusions of the study point to the fact that the main obstacles facing future disassembly and recycling technology are more economic than technological in nature (Boks, 1998). Similarly, automatic disassembly (for a limited product variety only) is expected to gain importance. Results for the automation of electrical and electronic consumer goods disassembly and recycling technology are presented in Table 3.4.

Automated disassembly is likely to be directly affected by the following factors (Boks, 1998):

- Product variety: Automated disassembly is likely to be more successful when independent of uniform products.

TABLE 3.4
Technical feasibility and economic attractiveness of automation of electrical and electronic goods disassembly technology

	Technical feasibility (%)			Commercial feasibility (%)		
Year	**Full**	**Partial**	**Limited**	**Full**	**Partial**	**Limited**
By 1998	2 (6)	7 (6)	29 (22)	4 (0)	2 (0)	12 (16)
By 2000	7 (2)	28 (14)	40 (35)	5 (0)	23 (4)	25 (16)
By 2005	26 (4)	39 (33)	24 (22)	11 (2)	25 (24)	39 (35)
By 2010	30 (28)	14 (29)	3 (18)	14 (15)	28 (31)	14 (16)
By 2015	11 (17)	11 (12)	2 (0)	23 (17)	5 (12)	2 (4)
By 2020	9 (15)	2 (4)	2 (2)	12 (17)	5 (10)	5 (4)
Later/never	16 (28)	0 (2)	0 (0)	32 (49)	14 (18)	4 (8)

Source: Adapted from Boks and Tempelman (1998).
Note: Feasibility for automotive products is indicated in parentheses.

TABLE 3.5
Obstacles to automating disassembly and their ranking

Obstacle	Year 2000	Year 2015
Large product variety	1	1
Number of similar products is too small	2–3	2
Product design is generally unfriendly for disassembly	2–3	3–5
General problem in return logistics	4–5	3–5
Variations in returned products	4–5	3–5
Highly damaged product	6–7	6–7
Damaged or corroded connections	6–7	6–7

Source: Adapted from Boks and Tempelman (1998).

- Only the recovery of light metals such as aluminum could make the disassembly of small household appliances attractive.
- Automated disassembly of hazardous substances is expected to be economically attractive.
- The value of material recovered will never be more than the cost of automated disassembly.
- Owing to technical innovations, demand for reusable parts is expected to be low. This implies that automated disassembly for separating reusable parts would be economically unattractive.
- An automobile disassembly time of 30 minutes would be possible only by using more destructive processes.

The main obstacles believed to prevent automated disassembly from becoming a commercially successful activity are enlisted in Table 3.5. Obstacles have been ranked from 1 (the most important) to 7 (the least important). The years 2000 and 2015 have been considered in this estimate.

3.6 USING PREDETERMINED MOTION TIME SYSTEMS (PMTS) TO EVALUATE EASE OF DISASSEMBLY: PROACTIVE APPROACH

Kroll and Carver (1999) developed design for disassembly metrics to be used when designing new products to make their disassembly for recycling easier. This method for quantifying the ease of disassembly of products was developed following several years of manual disassembly experiments on small electric appliances, computers and business equipment (Boks et al., 1996). Empirical data reveals that difficulty in product disassembly can be attributed to the following factors:

- Accessibility: Measure of ease with which a part can be reached by hand or by a tool.

- Positioning: The degree of precision required to place the tool or hand.
- Force: Measure of effort required to perform the task.
- Base time: The time required to do the basic task movements without difficulty. A task that poses accessibility problems, requires precise tool positioning or calls for exerting large forces takes a longer amount of time to complete. This time information has been converted to a simple linear scale in terms of difficulty ranging from 1 (the easiest) to 10 (the most difficult). Time spent in performing various actions was calculated using the MOST predetermined time system. If a disassembly task is divided into a series of elementary movements, the system can predict the time required by a worker with average skill to perform the task at an average pace. Rationalizing disassembly began with compiling two lists, tasks and tools, early in the process of experimenting with disassembly. The tasks list includes 16 standard disassembly operations such as unscrew, cut, wedge. The tools list comprises 24 tools such as wrenches, pliers, screw drivers and so on (Boks et al., 1996). The following methodology was adopted in the derivation of disassembly difficulty scores:
- Identification and definition of standard disassembly activities. Standard tasks were identified through observation of manual disassembly experiments. Slight variations in the performance of the same basic task were observed. The quickest and most efficient method for performing a task under average conditions was determined from the MOST system. This was designated as the standard task model.
- Investigation of various disassembly conditions: This included consideration of factors such as obstructions, handling difficulties and resistance. The effects of these conditions on performance time were assessed by assigning appropriate MOST parameter indices. Factors which complicated the disassembly received higher parameter indices thereby increasing overall task performance time. Parameter indices were reevaluated each time new conditions were imposed for a task.
- Categorization of sequence parameters: This was done according to the aspect of task performance measured by each parameter. The parameters were classified as follows: accessibility, positioning, force, base time and special. The category special was added in order to account for various nonstandard effects. Difficulty scores for Wedge/pry tasks are presented in Table 3.5. Difficulty scores for the "manual unscrew" and "power-assisted unscrew task" are presented in the Tables 3.7 and 3.8.

This methodology uses the so-called disassembly evaluation chart. The chart is presented in the form of two case studies in Tables 3.10 and 3.12. Each column heading in this chart is explained here:

- **Column 1—Part number**: The entry in column 1 stands for part number. This corresponds to the serial number of each part in the product. It is clearly recorded. A subassembly is counted as a part and a serial number with a suffix is assigned to it.

TABLE 3.6
Difficulty scores for disassembly task: "Wedge/Pry"

Wedge/Pry			Access	Position	Force	Base time
Slot allows easy insertion of pry bar	Clear	Light resistance	1	1	2	2
		Heavy resistance	1	1	2	2
	Obstructed	Light resistance	2	1	1	2
		Heavy resistance	2	1	2	2
Tight clearance, light pressure required to insert pry bar tip	Clear	Light resistance	2	1	1	2
		Heavy resistance	2	1	2	2
	Obstructed	Light resistance	3	1	1	2
		Heavy resistance	3	1	2	2
Tight clearance, stiff joint, heavy pressure required to insert pry bar tip	Clear	Light resistance	3	1	1	2
		Heavy resistance	3	1	2	2
	Obstructed	Light resistance	5	1	1	2
		Heavy resistance	5	1	2	2

Source: Adapted from Kroll and Carver (1999).

TABLE 3.7
Difficulty scores for disassembly task: "Manual unscrew"

Unscrew: Manual				Access	Position	Force	Base time
Single screw or nut		Clear	Light resistance	1	2	3	8
			Heavy resistance	1	2	10	8
		Obstructed	Light resistance	2	2	3	8
			Heavy resistance	2	2	10	8
Single bolt and nut	Clear bolt	Clear nut	Light resistance	1	3	3	8
			Heavy resistance	1	3	10	8
		Obstructed nut	Light resistance	2	3	3	8
			Heavy resistance	2	3	10	8
	Obstructed bolt	Clear nut	Light resistance	2	3	3	9
			Heavy resistance	2	3	10	9
		Obstructed nut	Light resistance	3	3	3	9
			Heavy resistance	3	3	10	9

Source: Adapted from Kroll (1996).

TABLE 3.8
Difficulty scores for disassembly task: "Power-assisted unscrew"

Unscrew: Power assisted			Access	Position	Force	Base time
Single screw or nut		Clear	1	2	1	3
		Obstructed	2	2	1	3
Single bolt with nut	Clear bolt	Clear nut	1	3	1	4
		Obstructed nut	2	3	1	4
	Obstructed bolt	Clear nut	2	3	1	4
		Obstructed nut	3	3	1	4

Source: Adapted from Boks et al. (1996).

- **Column 2—Theoretical minimum number of parts**: During the disassembly process, each part that is disassembled is evaluated to determine whether it is a required part in theory. If it is required as a separate component, a score of "1" is allocated to it. If it is not required as a separate component, a score of "0" is allocated to it. A subassembly that does not require to be disassembled any further is considered to be a single part.
- **Column 3—Number of repetitions**: The number of times a particular disassembly task needs to be performed is recorded. For instance, if three identical screws need to be unscrewed, the task "unscrew" is repeated three times.
- **Column 4—Type of task:** The research performed by Kroll considers 16 different types of tasks. Sometimes, it is necessary to resort to more than one task to accomplish a particular part disassembly. A "remove" task implies that a part can be gripped and removed with no additional motions.
- **Column 5—Direction of disassembly:** Direction of disassembly implies the axis along which the disassembly will take place. The XYZ coordinate system is affixed to the work bench upon which the disassembly is being performed with the positive Z axis pointing upwards. It is possible that a single task can be performed in multiple directions. The directions are listed in the order in which they occur.
- **Column 6—Tooling required for disassembly**: Kroll listed 28 common tools and used a combination of these tools for this research. The following tools were frequently used: Phillips screwdriver (PS), pliers (PL) and wire cutter (WC). Any disassembly operations that were carried out unassisted (by tools) were not noted separately. Tool manipulation (such as picking up a tool and putting it down) is implied by the presence of various tool codes.
- **Column 7–11—Rating of difficulty**: The difficulty with which each disassembly task can be performed is subjectively rated on a scale of 1–4 as follows:
  1: Disassembly task is performed without any difficulty.
  2: Disassembly task can be performed with some degree of difficulty.

3: Disassembly task can be performed with a moderate level of difficulty.
4: Disassembly task can be performed with a considerable amount of difficulty.

Accessibility: Accessibility describes the ease with which a part can be accessed. It is a measure of the amount of extant clearance and the ease with which a part can be manipulated during disassembly.

Positioning: Positioning describes the degree of precision with which a tool or the hand needs to be positioned and oriented to perform the task. For instance, when placing a screwdriver blade in the screw and fitting it in the screw slot, a higher level of accuracy is required. Contrast this with a task that requires a simple grip and removal operation.

Force: Force quantifies a measure of the total amount of force that must be applied in order to accomplish the disassembly task. For instance, a part that is press fitted is likely to require a greater amount of force to disassemble than a loosely fitted part. The same holds true for parts that have been glued.

Additional time: Additional time quantifies considerations that have not been otherwise accounted for. For example, a long screw is likely to take more time and vice versa.

Special: Any special problems encountered during the assembly process are listed in this section. These problems do not fit any of the other categories.

- **Column 12—Subtotal**: Subtotal is the sum of individual difficulty ratings for columns 7 through 11.
- **Column 13—Total**: This column contains the product of values entered in columns 3 and 12. It corresponds to the total difficulty rating of a disassembly task.
- **Column 14—Comments**: Any additional comments that would help in gaining a better understanding of the disassembly task are entered in this column. If any high ratings have been entered in columns 7–10, the corresponding reason is also entered in the comments section with the objective of helping to redesign the product. Any special tooling if required is mentioned in this space.

Overall disassembly efficiency of the product design is computed using the following formula:

$$\text{Overall Efficiency} = (5 \times \Sigma \text{ Column 2}) / (\Sigma \text{ Column 13}).$$

The overall efficiency can be construed as a percentage rating that describes the extent to which the current design is derived from a reference design of the same product. The reference design would contain the theoretical minimum number of parts with each part being disassembled with minimum effort.

Two case studies have been presented to more thoroughly explain this methodology. The first case study presents the disassembly operation of a handheld drill whereas the second case study deals with the disassembly operation of a computer central processing unit (CPU).

Case Study 1: Disassembly of a Handheld Drill

Consider the process of disassembling a handheld drill. The components of the drill are depicted in Table 3.9.

TABLE 3.9
A list of parts for a handheld drill

Part No.	Quantity	Part name	Component material
1	6	Front and rear screw	Steel
2	2	Middle screw	Steel
3	1	Upper housing	Plastic
4	–	Subassembly of parts 7–20	Mixed
5	2	Brush	Carbon
6	1	Lower housing	Plastic
7	1	Bushing	Bronze
8	1	Insulating washer	Phenolic
9	–	Subassembly of parts 18–20	Mixed
10	1	Rotor	Mixed
11	1	Spring washer	Steel
12	1	Washer	Steel
13	1	Bushing plate	Steel
14	1	Chuck assembly	Steel
15	1	Support plate	Steel
16	1	Shaft	Steel
17	1	Gear	Steel
18	1	Brushes wire	Mixed
19	1	Trigger assembly	Mixed
20	1	Stator	Mixed

Source: Adapted from Kroll and Carver (1999).

The disassembly evaluation chart for the disassembly process is presented in Table 3.10. The composition of the handheld drill is presented in Figure 3.4.

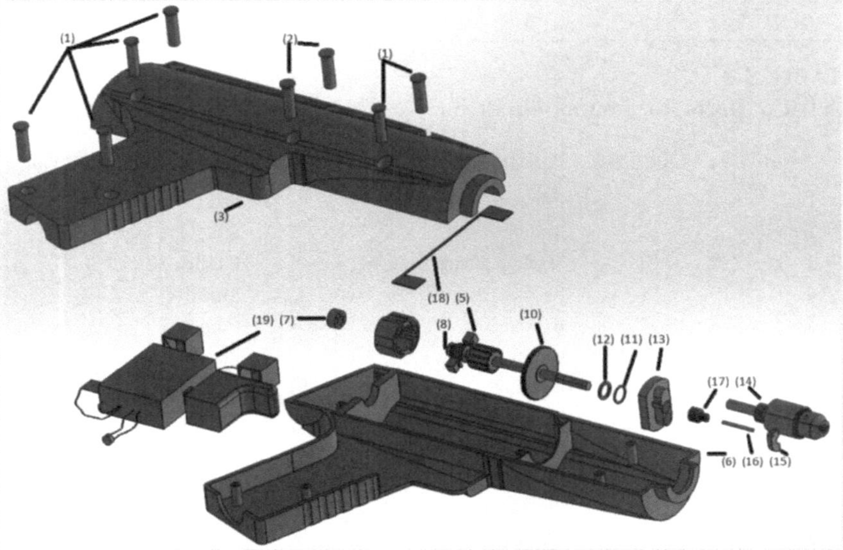

FIGURE 3.4 Schematic of a handheld electric drill (Adapted from Kroll and Carver 1999).

TABLE 3.10
Disassembly process for a handheld drill using disassembly time metrics derived from MOST

1	2	3	4	5	6	7	8	9	10	11	12	13	14
						Difficulty rating 1(Easy)–10 (Difficult)							
Part no.	Quantity	Minimum Number of Parts	Type of task	No. of task Repetitions	Required Tool	Accessibility	Positioning	Force	Base time	Special	Sub total	Total score	Comments
1	6	0	Un	6	PS	1	2	1	3	1	8	48	Magnetic power screwdriver is used
2	2	0	Un	2	PS	2	2	1	3	1	9	18	Magnetic power screwdriver is used
3	1	1	We	3	PB	2	1	1	2	1	7	21	Pry off at three locations
			Fl	1		1	1	1	2	1	6	6	Flipped and held for cleaning
			Cl	1	RG	2	1	1	2	1	7	7	Lubricating grease is cleaned
4 (sub)		0	We	2	PB	1	1	1	2	1	6	12	Sub is pried out of lower housing
			Re	1		2	1	1	1	1	6	6	Bulky wires tangle
5	2	2	Re	2		1	1	1	1	1	5	10	Pop out upon removing sub 4

(continued)

TABLE 3.10
(Cont.)

1	2	3	4	5	6	7	8	9	10	11	12	13	14
						Difficulty rating 1(Easy)–10 (Difficult)							
Part no.	**Quantity**	**Minimum Number of Parts**	**Type of task**	**No. of task Repetitions**	**Required Tool**	**Accessibility**	**Positioning**	**Force**	**Base time**	**Special**	**Sub total**	**Total score**	**Comments**
6	1	1	Ho	1		1	1	1	2	1	6	6	Hold for cleaning and toss in bin
			Cl	1	RG	2	1	1	2	1	7	7	Lubricating grease to be cleaned
Disassembling subassembly 4													
7	1	1	Pu	1		1	1	1	1	1	5	5	
8	1	1	Pu	1		1	1	1	1	1	5	5	
9 (sub)		0	Pu	1		1	1	1	1	1	5	5	Sub 9 is put aside
10	1	1	Pu	1		1	1	1	1	1	5	5	
			Cl	1	RG	1	1	1	2	1	6	6	Clean lubricating grease
11	1	1	Pu	1		1	1	1	1	1	5	5	
12	1	0	Pu	1		2	1	1	1	1	6	6	
13	1	1	Pu	1		1	1	1	1	1	5	5	

14	1	1	Re	1		1	1	1	1	1	5	5	
15	1	1	Pu	1		1	1	1	1	1	5	5	
16	1	0	Pu	1		1	1	1	1	1	5	5	Pull shaft from gear 17 and toss tooth
Disassembly of subassembly 9													
18	1	1	We	2	ST	2	1	1	2	1	7	14	Pry cable shoes with a pointed punch
			Pu	2		1	1	1	1	1	5	10	Pull wire from stator (part 20)
19	1	1	We	2	ST	2	1	1	2	1	7	14	Pry cable shoes with a pointed punch
			Pu	2		1	1	1	1	1	5	10	Pull wire from stator and toss tooth

Source: Adapted from Kroll and Carver (1999).

Case Study 2: Disassembly of a Computer CPU

A case study of disassembling a computer CPU is presented here.

Table 3.11 depicts the components of a typical computer CPU. The figures for individual parts and subassemblies for the CPU are presented in Table 3.12.

TABLE 3.11
A list of parts for a computer CPU

Part no.	Quantity	Part name	Component material
1	4	Screw for motherboard	Steel
2	4	Drives subassembly screw	Steel
3	5	Fastener for motherboard	Plastic
4	4	Foot	Rubber
5	2	Screw for printer port	Steel
6	4	Power supply screw	Steel
7	1	Screw for speaker	Steel
8	1	Speaker	Mixed
9	2	Hard drive cover panel screws	Steel
10	1	Hard drive cover panel	Plastic
11	–	Drives subassembly parts 23–29	–
12	1	Power supply	Mixed
13	4	Port screw	Steel
14	1	I/O Card #1	Mixed
15	1	I/O Card #2	Mixed
16	1	I/O Card #3	Mixed
17	1	Motherboard	Mixed
18	4	Top screw on front cover	Steel
19	2	Bottom screw on front cover	Steel
20	1	Front bezel	Plastic
21	1	Filler panel	Plastic
22	1	Housing	Steel
23	6	Drives subassembly side screw	Steel
24	4	Drives subassembly bottom screw	Steel
25	1	Drives subassembly frame	Steel
26	1	Hard drive	Mixed
27	4	Floppy disk drive screw	Steel
28	1	Connector plat	Steel
29	2	Floppy disk drive	Mixed

Source: Adapted from Kroll and Hanft (1998).

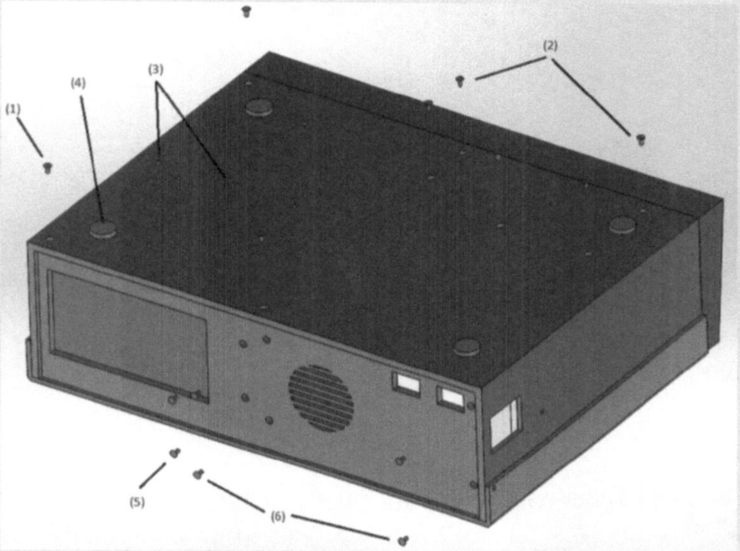

FIGURE 3.5 CPU placed in an initial upside-down position (Adapted from Kroll and Hanft, 1998).

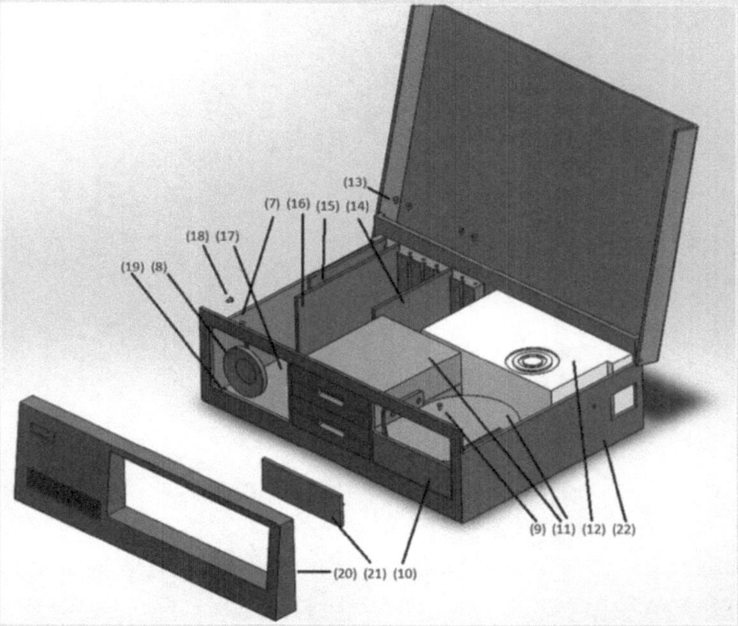

FIGURE 3.6 Components of the CPU without cables and depicting all individual components (Adapted from Kroll and Hanft, 1998).

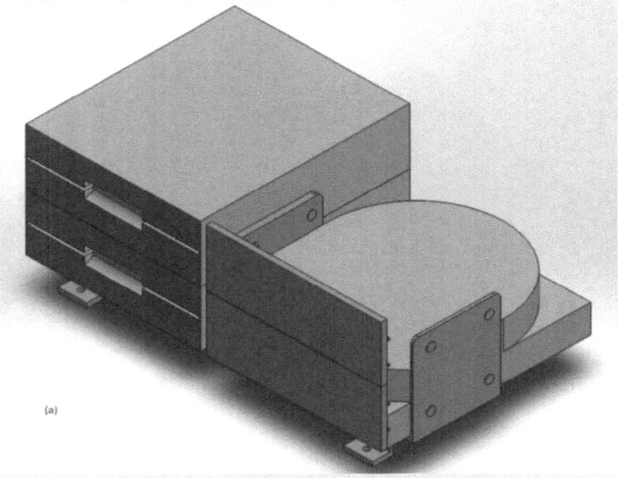

FIGURE 3.7A Drives' subassembly in the assembled state (Adapted from Kroll and Hanft, 1998).

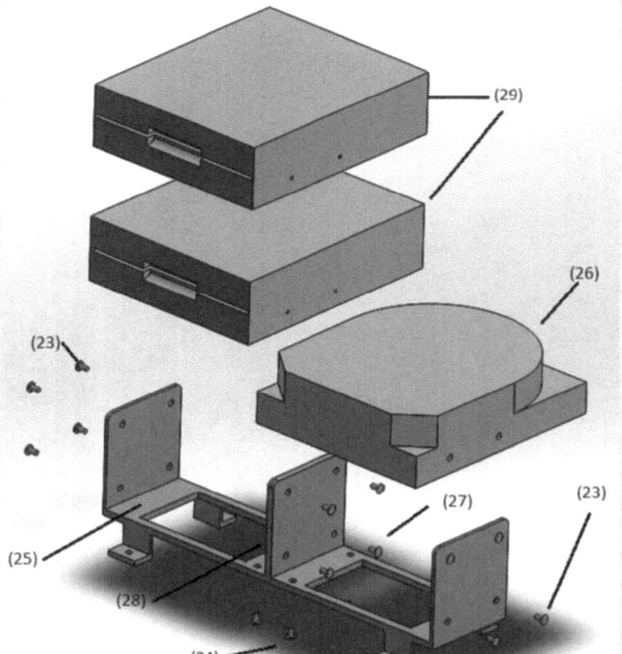

FIGURE 3.7B Drives' subassembly: Exploded view (Adapted from Kroll and Hanft, 1998).

TABLE 3.12
Disassembly process for a computer CPU using disassembly time metrics derived from MOST

1	2	3	4	5	6	7	8	9	10	11	12	13	14
						Difficulty rating 1(Easy)–10 (Difficult)							
Part no.	Quantity of parts	Minimum number of parts	Type of task	No. of task repetitions	Required tool	Accessibility	Positioning	Force	Base time	Special	Subtotal of scores	Total score	Comments
1	4	0	Un	4	PS	1	2	1	3	1	8	32	
2	4	0	Un	4	PS	1	2	1	3	1	8	32	
3	5	0	Cu	5	SH	2	2	1	2	1	8	40	Head of plastic fasteners is cut off with shears
4	4	1	Pe	4		1	1	1	2	1	6	24	
5	2	0	Un	2	ND	1	2	1	3	1	8	16	
6	4	0	Un	4	PS	1	2	1	3	1	8	32	
7	1	0	Fl	1		1	1	3	2	1	8	8	Assembly is flipped right side up
			Pu	1		1	1	1	1	3	7	7	Top is pulled open (> 12") while pressing buttons
			Un	1	PS	1	2	1	3	1	8	8	
8	1	1	Pu	1		1	1	1	1	1	5	5	Speaker plug is disconnected from motherboard

(continued)

TABLE 3.12
(Cont.)

1	2	3	4	5	6	7	8	9	10	11	12	13	14
						Difficulty rating 1(Easy)–10 (Difficult)							
Part no.	Quantity of parts	Minimum number of parts	Type of task	No. of task repetitions	Required tool	Accessibility	Positioning	Force	Base time	Special	Subtotal of scores	Total score	Comments
			Re	1		2	1	1	1	1	6	6	Connector tangles in hole when speaker is removed
9	2	0	Un	2	FS	2	2	1	3	1	9	18	Use magnetic tip screwdriver
10	1	1	Re	1		1	1	1	1	1	5	5	
11		0	Pu	1		2	1	2	1	6	12	12	Pull floppy plug. Requires 6 actions, obstructed access, resistance)
(sub)			Ho	1		3	1	1	2	1	8	8	Bulky drives subassembly frame is gripped.
			Pu	3	PL	2	2	2	2	3	11	33	Pull 3 power plugs (3 actions, obstruction, resistance)
			Pu	1		1	1	2	1	1	6	6	Hard drive plug #1 is pulled (clear, resistance)
			Pu	1		2	1	2	1	1	7	7	Hard drive plug #2 is pulled (obstruction, resistance)
			Pu	1		2	1	2	1	2	8	8	Drives subassembly is pulled (2 actions, obstruction, resistance)

12	1	1	Pu	1		2	1	1	1	1	6	6	Move to make room (obstruction, light resistance)
			Pu	2		1	1	2	1	1	6	12	Disconnect plug from motherboard (clear, resistance)
			Re	1		1	1	1	1	1	5	5	
13	4	0	Un	4	PS	1	2	1	3	1	8	32	
14	1	1	Pu	1		1	1	2	1	1	6	6	
15	1	1	Pu	1		1	1	2	1	1	6	6	
16	1	1	Pu	1		1	1	2	1	1	6	6	
17	1	1	Re	1		1	1	1	1	1	5	5	
18	4	0	Un	4	PS	1	2	1	3	1	8	32	
19	2	0	Un	2	PS	2	2	1	3	1	9	19	Access to part is obstructed
20	1	1	Re	1		1	1	1	1	1	5	5	
21	1	1	Re	1		2	1	1	1	1	6	6	Access to part is obstructed
22	1	1	Re	1		1	1	2	1	1	6	6	Part is heavy
Disassembling subassembly 11													
23	6	0	Un	6	PS	1	2	1	3	1	8	48	
24	4	0	Fl	1		1	1	3	2	1	8	8	Component is heavy
			Un	4	PS	1	2	1	3	1	8	32	
25	1	1	Re	1		1	1	1	1	1	5	5	
26	1	1	Re	1		1	1	1	1	1	5	5	
27	4	0	Un	4	PS	1	2	1	3	1	8	32	
28	1	0	Re	1		1	1	1	1	1	5	5	
29	2	2	Re	2		1	1	1	1	1	5	10	

Source: Adapted from Kroll and Hanft (1998).

3.7 FRAMEWORK OF AN INTERACTIVE SYSTEM FOR DESIGN FOR DISASSEMBLY

This section of the chapter presents an interactive design for disassembly system. Thus far, in this chapter, we have studied a host of design for disassembly guidelines, some economic models, even a methodology based on PMTS. The system presented here is meant to be an addition to conventional CAD systems. Its objective is to aid the designer in making quicker and more efficient design decisions without needing to refer to the literature.

When we say "interactive design system," we mean that the design system and the developer interact with each other. This interaction takes place through different modes of communication that elicit a design response from the designer. The communication takes place through a series of questions. Designer's responses to these questions enable the system to structure a coherent response with the objective of solving the design problem.

In other words, the system asks the designer specific questions with respect to the design and solicits responses. The responses, in turn, are used to further simplify the task of the designer. This simplification is achieved by means of the system offering a variety of design solutions.

As we have already studied in this chapter, disassembly can be defined as the process of taking apart a product. A product consists of an assembly of components. The different components of the products are held together by a variety of fasteners. As its name implies, a fastener is used to hold together two disparate components. Components could be individual parts or could be in the form of subassemblies. In other instances, a fastener could also serve other functions such as transmission of force. This could be its secondary function and depends upon the nature and architecture of the product.

Whenever design for disassembly principles is used in product design, they will almost always result in the use of a fastening technique that satisfies its main purpose (namely fastening) but is simultaneously easy to take apart. This objective is satisfied by using the many design principles for facilitating disassembly that we have already considered in this chapter.

Examples of fastening techniques that are chosen in line with the aforementioned design principles includes using snap fits, standard types of fasteners or minimizing the number of fasteners. Different parameters that affect the process of choosing an appropriate fastener are depicted in Figure 3.8.

The architecture of the interactive design system also includes so called design libraries. These are contained within the disassembly module. The function of the design libraries is to help the designer to select appropriate design features. If the system were to prompt the designer to enter the type of fastener that would serve the intended function, the designer is free to select from a variety of fasteners (such as snap fit, screw in fastener, rivets, etc.). These choices are then ranked in order of preference. Doing so enables the system to use that feedback into the design solution.

The different types of screw in fits, rivets, snap fits and so on and their pertinent information is contained in separate design libraries. This information is used and recalled when prompted to do so by the user. The structure of such design libraries is presented in Figure 3.9. Similar design libraries can be used for design features such as weight, shape, material, surface finish, tools and so on.

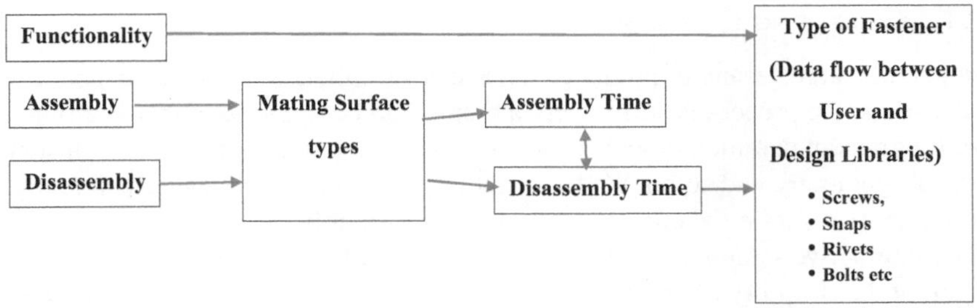

FIGURE 3.8 Process of selecting a fastener for disassembly while maintaining functionality.

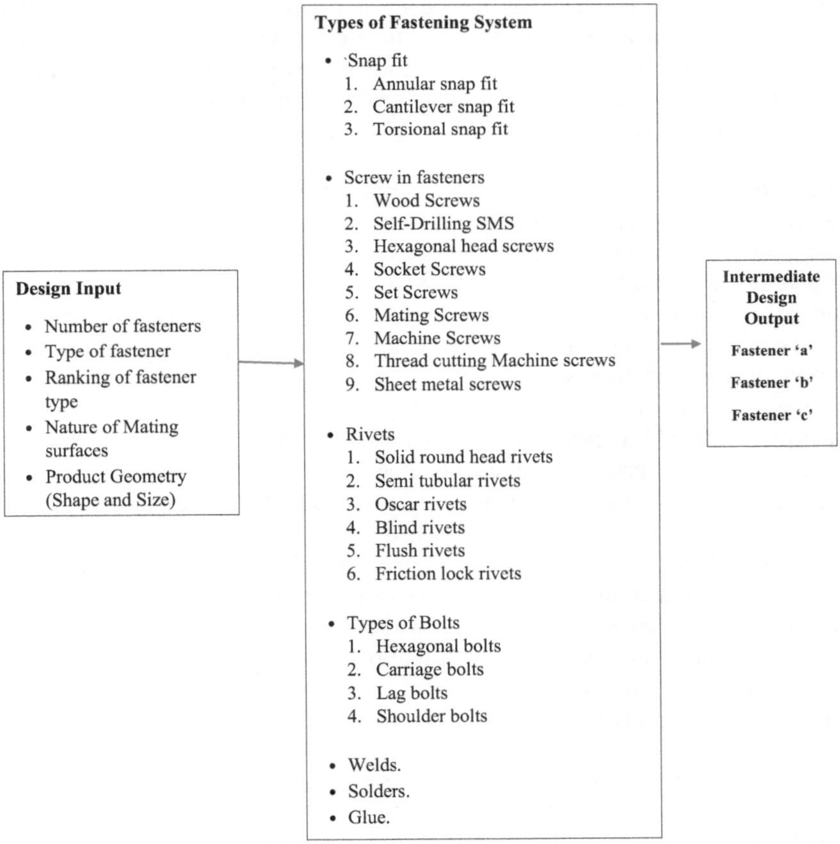

FIGURE 3.9 Design library depicting types of fasteners used for assembly and disassembly processes.

3.7.1 Computing Disassembly Time

Owing to the large variety of products that a disassembly facility needs to handle on a regular basis, the process is still largely a manual process. Disassembly is a function of various design parameters such as weight, location, number and types of fasteners used, part geometry and so on. We have studied this principle repeatedly not only on this chapter but also in the chapter on design for assembly.

The interactive system presented here would enable the designer to compute the amount of disassembly time for a specific disassembly operation. This computation is performed using information from the motion time measurement (MTM) system. We have studied an example of this system in the chapter on design for assembly. Process disassembly time is computed by simply adding together the disassembly times for each step. This enables the designer to arrive at the total disassembly time for the entire process. It is necessary to add the amount of setup time to this disassembly time. The design database in integrated together with the feature that computes the disassembly time. Doing so enables every design option offered up by the system to also include an additional piece of data, namely "disassembly time" with it.

Next, disassembly cost can easily be computed from disassembly time. This is achieved by multiplying disassembly time with standard direct labor rates. This particular feature further demonstrates the practical utility of the system. This feature as a part of the overall system is depicted in Figure 3.11. It is a continuation of Figure 3.10. After having chosen the best type of fastener for a particular application, the disassembly time for that fastener type is then computed by the system and the corresponding information is output.

It is clear from what has been said thus far that the total disassembly time is uniquely determined for each set of design parameters. This implies that if the design parameters were changed, the disassembly time would change as well. Before product design changes are finalized, a mutually beneficial tradeoff between functionality and ease of disassembly would need to be reached.

Take for instance the case of a washing machine. A washing machine drum subassembly is heavy and unwieldy. It consists of several moving parts and is secured by a combination of screws, interlocking switches, suspension rod assemblies and so on and other constraints and fasteners. It will be more difficult and time consuming to disassemble such a subassembly than disassembling a printed circuit board (PCB) that is constrained and held in place by means of two snap fits.

On the contrary, it is likely that the same drum subassembly will be easier to disassemble if snap fits were holding it in place instead of screws. The functional aspect of the fastening process would be compromised by such a scenario. That four snap fits would be able to successfully constrain the subassembly in place especially given the weight, size and shape of the components is highly doubtful. Thus, it is important to successfully resolve the conflict between functionality and design for disassembly prior to making any design changes final.

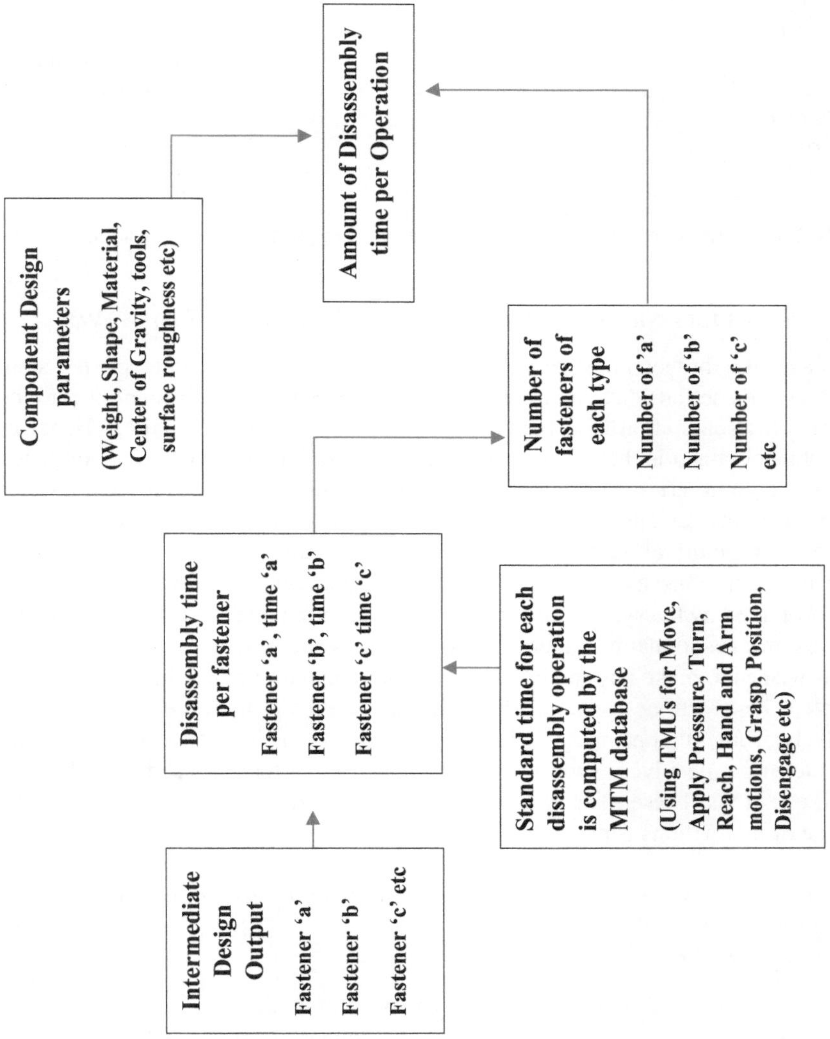

FIGURE 3.10 Computation of total disassembly time for each design configuration.

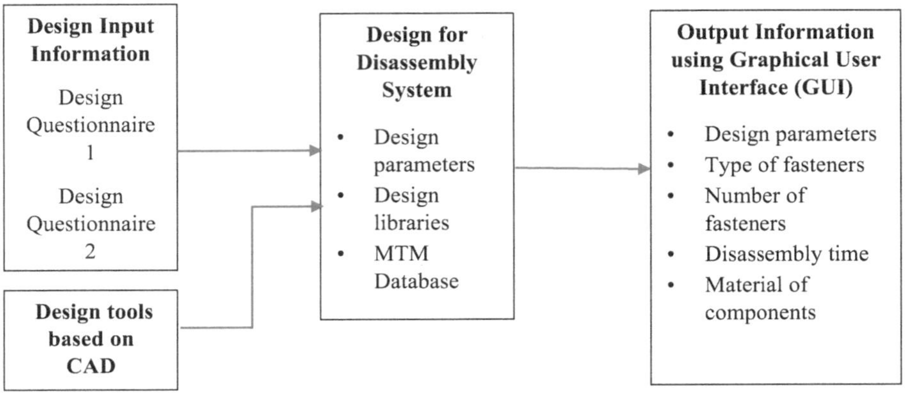

FIGURE 3.11 Overall configuration of the general design for disassembly system.

3.7.2 How Does the Interactive Design for Disassembly System Work?

As we have already seen in the previous section of this chapter, the interactive system being discussed solicits design information and other pertinent information from the designer. This solicitation is in the form of an interactive questionnaire. The information thus obtained is then synthesized and processed together with data obtained from the design libraries. The goal is to provide a possible design solution or a series of solutions to the designer so that the designer can pick the one that is most suitable. The general structure of the system is presented in Figure 3.11.

The first step consists of the user completing the component design. Next, the user is prompted by the system through detailed questionnaires, design questionnaire 1 and design questionnaire 2. During this step, the system tries to find out whether the user wishes to make any modifications to their design to make them easy to disassemble. The questions contained in the questionnaires prompt the user to answer objectively. Thus, the only way to answer these questions is to answer: Yes or No. Upon eliciting a positive response from the user, the system proceeds to ask more detailed questions. These "Tier 2" questions would result in numeric responses. This sequence of interactions is presented here:

> Question 1: Do you wish to choose a specific feature that can be altered for design for disassembly?
> User response: Yes.
>
> Question 2: How many components do you wish to modify?
> User response: 1/2/3/4…m" (This results in a numeric response, where "m" is the final number of components).A sample list of probable interactive yes/no type of questions (design questionnaire 1) for design modification for disassembly is presented in Table 3.13.

The interactive system is built on the basic premise of Design for X, where "X" can represent a variety of design objectives ranging from manufacturing and

TABLE 3.13
Preliminary questions enable design for disassembly system to interact with user

Interactive questions for design for disassembly	Probable response
Do you wish to modify product design for ease of disassembly?	Y/N
Do you wish to choose any specific component to be modified for ease of disassembly?	Y/N
Do you wish to add any constraints within which the assembly process can occur for the component chosen?	Y/N
Do you wish to specify the material of the component(s)?	Y/N
Do you wish to change the spatial alignment of the component(s)?	Y/N
Do you wish to wish to make any changes to the component design by adding design features for ease of disassembly?	Y/N
Do you wish to make changes to the component design by subtracting any design features for ease of disassembly?	Y/N
Do you wish to combine two or more components to facilitate disassembly?	Y/N
Do you wish to change the component material to facilitate disassembly?	Y/N
Do you wish to modify the fastening system to facilitate disassembly?	Y/N
Do you wish to modify component design by using standardized shapes and/or features to facilitate disassembly?	Y/N
Do you wish to make any changes to component design to enable ease of handling for disassembly?	Y/N
Do you wish to make any changes to component design to enable ease of insertion for disassembly?	Y/N

assembly to disassembly, functionality and usability. The structure of the interactive system presented thus far constitutes the first part of the process of design modification for X (in this case, X represents the design goal of disassembly). If the user responds to any question in the negative (e.g., Do you wish any specific component to be modified for ease of disassembly? N), it has the effect of the system moving on to the next question. This sequence is repeated until all questions have been answered.

The decision-making loop is stopped if the system encounters a negative response to any question. On the contrary, a positive response to a question makes the system switch to the second step of the design modification for X process. The second step requires the user to provide information in the form of design data. Being that most design data are objective and are represented by numbers, a change in spatial arrangement of components or combination of two or more different components would prompt the system to ask pertinent design questions. An additional set of questions is asked by the system to provide further directions. These questions are based on user response to the questions in step 1. An example of such questions and directions is presented in Table 3.14.

The system being described in these pages is intended to exist as a standalone system and can be appended to standard CAD packages available on the market. System attributes together with detailed questionnaires can, however, be embedded into CAD packages. The main goal of this is to prompt a response from the designer.

TABLE 3.14
Level 2 questions gather specific information pertinent to ease of disassembly

Interactive questions/directions for design for disassembly	System direction
Please specify material of the component	Direct to material list
Please specify fastening system to be used	Direct to fastening systems
Please choose specific component(s) to be modified for ease of disassembly	Highlight specific components
Please choose components you wish to combine for ease of disassembly	Highlight specific components
Please choose handling method for ease of disassembly	Direct to handling method list
Please choose specific components whose spatial alignment needs to be changed	Highlight specific components
Please specify new dimensions on the components you wish to modify	Direct to dimensions
Please specify the standard set of fasteners you wish to use	Direct to fastening systems
Please specify the design features you wish to add for disassembly	Direct to CAD system
Please specify the design features you wish to subtract for disassembly	Direct to CAD system
Please choose standard design features you wish to include for ease of disassembly	Direct to standard design features list
Please specify the value of surface finish you wish to use on the surface	Direct to standard surface finish list or customize
Please specify the particular disassembly method to be used	Direct to nondestructive disassembly methods list
The material you chose is included in the EPA's list of hazardous materials. Do wish to choose another material?	Direct to material list
Please select the tools you wish to use for disassembly	Direct to tools list
The assembly joints you chose are inaccessible for disassembly. Do you wish to change them?	Highlight joint(s)
Please select standardized alignment features for ease of disassembly	Direct to standard design features list

This response is provided at crucial steps during the design process. It pertains to component/product design modification for assembly, disassembly or maintenance. This feature can be built upon and expanded in the course of future research.

3.7.3 Case Study

Thus far, we have presented the overall structure on an innovative interactive design system that can be used to modify product design. The system can be used for any number of design goals. The design goal being dealt with in this chapter is disassembly.

In order to gain a better understanding of this system conceptually, we will present a case study of a real product in this section. The design of the product will be altered to facilitate disassembly. Design decisions will be made using the system described in this chapter. Doing so will help further demonstrate the utility value of this system.

TABLE 3.15
Bottom hatch and access panel of a typical laptop computer: Different components

Component	Quantity	Material
Bottom panel	1	Aluminium
Access hatch	1	ABS plastic
Fastening screws	5	Aluminium
Laptop main body	1	Various

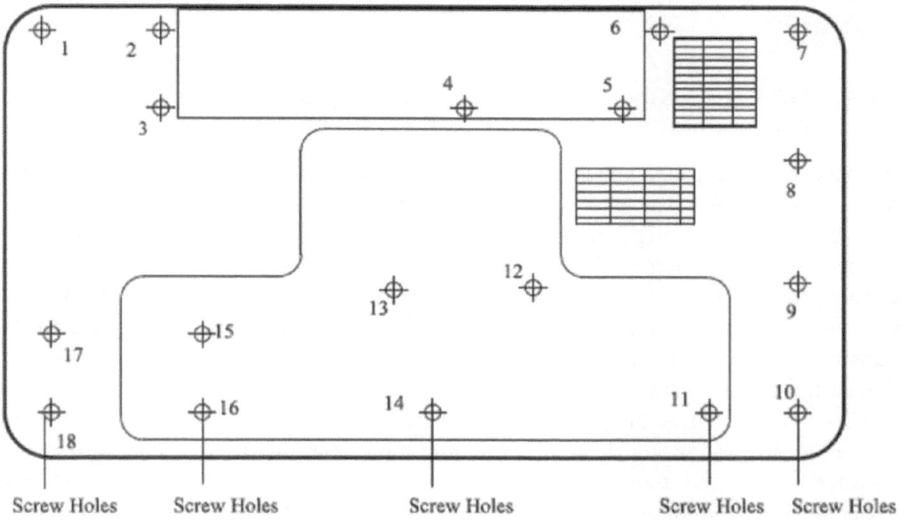

FIGURE 3.12 Bottom panel of a typical laptop computer.

Let us take the case of disassembling a laptop computer for our case study. As a result of an ever-increasing need for "computing," computers are found everywhere. Computers are so commonplace that the trend is increasingly shifting towards mobile computing. Consequently, desktops are being replaced by laptop computers in large numbers. This is the reason we picked this particular product to help demonstrate the utility value of our interactive system. It is possible for a designer to alter and modify different parts' design of the laptop computer. Once again, the goal of this case study is to aid in disassembling the laptop computer by using the methodology presented in this chapter.

Design objective: *To alert and modify the design of the bottom hatch of a laptop computer. The goal is to improve access to internal components, thereby improving ease of disassembly.*

Consider Figure 3.12. The bottom panel of a typical laptop computer is depicted in this figure. The goal of the designer is to alter and modify the design of the access hatch so that access to internal components is facilitated. This will, in turn, improve the overall ease with which the laptop can be disassembled. The access hatch is usually situated in the bottom panel and serves the following purposes:

- Both memory modules are protected by the hatch.
- The hard drive is secured and protected by the hatch.
- The CMOS battery is secured and protected by the hatch.
- The wireless card is secured and protected by the hatch.

The access hatch of a laptop computer is not to be taken lightly. It is an obviously important component in view of its functionality as described earlier. As a result, it is important to design it carefully. When the design of the access hatch is modified, it will result in affecting the design of other components as well. They are presented in Table 3.15.

3.7.4 How is a Product Being Designed Based on Current Practice and Literature?

We have presented an exhaustive overview of design for disassembly guidelines in this chapter. We also carefully studied the literature on the topic. This study has revealed some interesting facts. We found that most disassembly related research has focused on process logistics such as product takeback and optimizing the disassembly process. It also deals with other issues such as balancing the disassembly line, ways and means to achieve maximum profit and formulation of disassembly sequence plans and so on. We have looked at the basic definition of a disassembly sequence plan earlier in this chapter itself.

These are strategies that are largely reactive in nature. As such, one gives consideration only to current design, not design modifications. Design alterations and modifications in line with design for disassembly principles are not considered. No proactive design strategies are investigated. This is what is referred to as process optimization and it does not factor in design considerations. If there are any proactive design decisions being incorporated into the design process, they only address the twin objectives of manufacturability and assembly. Also, we have already studied that not too many proactive design for disassembly tools are available in the market today.

One of the proactive design for disassembly tools we discussed in this chapter is the method presented by Kroll and Carver (1999). This method seeks to quantify disassembly effort. It is achieved by rendering it as a function of underlying design parameters. One of the drawbacks of this method is that it does not offer any means by which the component in question could be redesigned for ease of disassembly. Thus, insofar as its capacity of being able to offer design solutions, the method is found lacking. In addition, it was found to not be very user friendly because it does not easily lend itself to automation.

Consider the design guidelines formulated by Zhang et al. (1997) to enable design for disassembly. Once again, these are just guidelines, and a comprehensive methodology was found lacking. Similar guidelines were also developed by Suga et al. (1996) by formulating quantitative disassembly evaluation criteria. The goal was to evaluate a design from the perspective of rendering the product easier to disassemble.

These guidelines would enable assessment of the current design or any future designs for ease of disassembly. Numeric scores based on the quantitative criteria could be assigned to each design in order to enable design assessment. But the methodology fell short of explaining how the design modifications were to be achieved. Given the lack of a comprehensive methodology, no practically applicable design tool was proposed.

A methodology to enable automatic generation of disassembly sequences was developed by Ong (1999). Once again, as in numerous prior examples, the methodology sought to generate disassembly sequences was based on the current design. No information was provided on how to modify the existing design so that it would be easy to disassemble.

As far as we know, no comprehensive design tool is currently in existence that seeks to bring together the myriad guidelines presented thus far in this chapter. The format needs to be coherent in structure so that the information can be offered in the form of an interactive design solution to the user. The framework being presented in this chapter tries to achieve that goal.

It is intended that the interactive system will achieve design goals quickly, thus simultaneously enhancing productivity. Given the high degree of automation inherent in the process, the system will result in reducing the amount of effort and time required to achieve modifications. This will be in large part due to the way it "communicates" with existing CAD packages. System productivity can be enhanced manifold if design renderings take on the form of 3D objects instead of 2D renderings. This is because it is generally easier to visualize a design in three dimensions than in two. Doing so facilitates both component design as well as the spatial arrangement between other components and subassemblies.

One of the main characteristics of the proposed framework is that it offers multiple design solutions. Using these solutions, it would be possible for the designer to visualize and imagine a host of different design architectures. The more solutions one has, the more it is possible to use a variety of permutations and combinations to achieve better end designs. Design time is likely to be minimized given the fact that the proposed system provides solutions in a visual format. A combination of these features makes the design process less irksome.

When compared to extant design software, it will be observed that disassembly design solutions rendered by standalone CAD software using CAD packages is often inefficient and consumes too much time. The reason for this is that the designer needs to build each solution from scratch. This in turn limits the time, energy and drive of the designer, whereas in the present case, solutions are proposed by the system. As such, they are easier to visualize and can be built upon quickly. This has the effect of accelerating the design process.

3.7.5 Solution in the Form of Design Modifications

The solution proposed by the interactive system will achieve design modifications by adhering to the following steps:

1 During the first step, the CAD design of the access hatch and the bottom panel is rendered on screen.

2 During the second step, the user is prompted by the design for disassembly system to indicate whether they would like to modify the design to make it easy to disassemble. This is achieved by using level 1 questions as described in Table 3.13.
3 Step three depends on the user input (either Y or N). Upon receiving an answer in the affirmative (Y) to the question, the system would proceed to the next question. At this stage, the system would prompt the user to specify the part of the design they would like to modify (such as material, fastener type, fastener location, etc.).
4 As step four, it would be indicated by the designer that they would like to modify the design of not only the bottom panel but the access hatch as well.
5 Figure 3.13 indicates the five screws to secure the access panel to the bottom hatch.
6 All information that is required in order to make the desired changes is acquired by the system by asking specific questions of the designer. These are typically level 2 questions. Examples of these questions are presented in Table 3.14. After the designer has provided the information that was requested of them, the resulting system output is in the form of the myriad design options that the designer can pick and choose from in order to make an informed design decision. Examples of such design options include the variety of materials, the different types of fasteners that can be chosen such as screws, snap fits, rivets and so on. The different amount of disassembly time for different fastener types is provided by the system and is computed using MTM data. The design libraries (which have already been discussed) provide information on the various design options.

A likely sequence of communication is provided as follows:

Q.1: Do you wish a particular component that can be modified for ease of disassembly? (Level 1 question)
Response: Y.

Q.2: Please choose any specific components to be modified for ease of disassembly (Level 2 question)
Response: Upon being asked this question, the designer would highlight particular CAD renderings of the access hatch and bottom panel.

Q.3: Do you wish to modify the fastening system so that disassembly can be facilitated?
Response: Y

Q.4: Please use standard fastening method(s) you wish to use (Level 2).
Response: Designer chooses standard screws and snap fits.

Q.5: Please specify specific disassembly method to be used (Level 2).
Response: Designer chooses “unscrew” and “pry.”

Q.6: Please use standardized alignment features for ease of disassembly (Level 2).
Response: Designer chooses mating surfaces as depicted in Figure 3.12.

7 After the system has received all the relevant information as collected by the information exchange described, the system is then likely to suggest some specific design solutions as follows. The system solves the design problem by using the information stored in the respective databases (as we have pointed out in our discussion thus far).

3.7.6 Design Option I

One of the first design options could consist of maintaining the design parameters as they exist. This includes parameters such as shape and material of the respective panels. The system could propose that the number of fastener screws could be reduced to four. This is accomplished by eliminating the screw in the center of the panel. Doing so would reduce the total disassembly time by as much as 20%.

3.7.7 Design Option II

The second design option suggested by the system could center around altering geometry of the part. It could suggest that the shape of the access hatch could be modified to a trapezoid instead of a rectangle. Also, it could recommend that three screws be used at each corner to fasten the hatch. As far as material of the panel is concerned, it would be unchanged. It is expected that doing so would reduce the disassembly time by as much as 25%.

3.7.8 Design Option III

The third design option suggested by the system could consist of keeping the shapes of the panels unchanged. However, the fastening method could be changed from screws to snap fits. Refer to Figure 3.13: The snap fits would be placed at the edges highlighted in the figure. Several goals are achieved simultaneously by making this design modification. They are listed as follows:

- Both product assembly as well as disassembly processes can be accomplished easily.
- The panel and access hatch will continue to use the same material.
- The parts will retain their functionality as well as their appearance. As a matter of fact, it will lead to an enhancement in the appearance because surface features such as holes will be eliminated.
- The surface will be smoother in appearance and no longer resemble the one that is riddled with bullets (due to the holes). The new surface will retain its functionality and the assembly and disassembly processes will take less time.

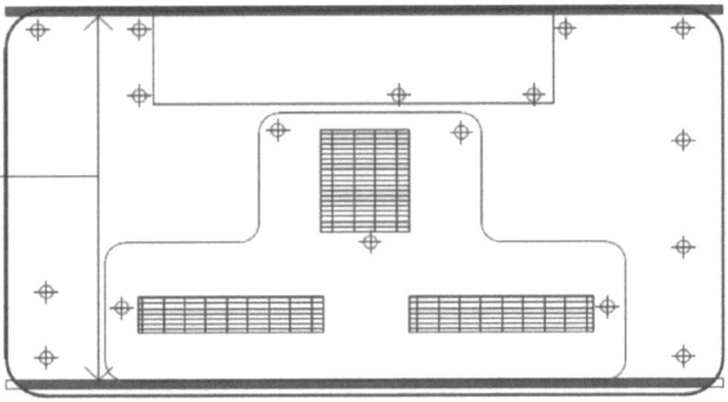

FIGURE 3.13 Placement of snap fits is shown by design option III.

- The only additional consideration confronting the designer would be the added cost that is necessary to incorporate the suggested design changes. This can be construed to be more a policy matter.

The design of the bottom panel can be modified by having a similar exchange of information between the system and the designer.

3.7.9 Advantages of the Proposed System

The advantages of the system can be enumerated as follows:

1 The system will act as an add-on to existing CAD packages. As a result, it will support CAD packages and need not be set up independently.
2 The system being described in this chapter centers on product design and thus product optimization, not process optimization. Design features that hamper ease of disassembly will not be incorporated into the product design to start with. We already know that over 80–90% of disassembly related gains is attributable to product design. This implies that only 10–20% can be attributed to process optimization of the disassembly process. Focusing on product design will reduce additional cost that the manufacturer would have to bear to dispose of the product in an environment-friendly manner.
3 It has been found that ease of disassembly also leads to ease of maintainability. This implies that the design changes recommended by the system are likely to result in a product that is easy to maintain as well. From the several options offered by the system to the designer, the latter is free to choose the option that best suits their needs. In addition, cost is another constraint that needs to be considered when making the final decision. The system will offer the designer several design options from which the latter is free to choose the most feasible one. It should be noted that cost plays a crucial role in making the final decision. Thus, the designer may not always be free to choose the option that is most "disassembly friendly." Nevertheless, in light of the fact

that more than one option is available, the option that is chosen in the end is likely to be more disassembly friendly than the current design.

4 Visualization of the design solutions is rendered easy by virtue of the fact that the system utilizes a graphical user interface (GUI) to display the various designs on screen. It also relies upon solid models to facilitate quick and exhaustive design comprehension. Using this combination allows better and more in-depth understanding of the design. This eventually is likely to lead to a solution that is better than the one conceptualized by the designer.

3.8 CONCLUSION

In this chapter, we investigated the role of product disassembly as one of the crucial aspects of designing sustainable products. Disassembly is important so that valuable material from end-of-life products can be recovered. It also plays an important role in recovering components and subassemblies that can be reused in future products. To this end, several design for disassembly guidelines were presented. At least three different case studies were also presented.

REFERENCES

Benson, A. (1996). "Engineering cars for recycling," in: Assembly, pp: 42–45.

Billatos, S.B. and Nevrekar, V.V. (1994). "Challenges and practical solutions to designing for the environment," ASME Design for Manufacturability Conference, Chicago, pp: 49–64.

Boks C. and Templemon, E. (1998), "Future disassembly and recycling technology. Results of a Delphi study," Futures, Vol: 30, No: 5, pp: 425–442.

Boks, C., Kroll, B.E., Brouwers, W.C.J. and Stevels, A.L.N. (1996). "Disassembly modeling: Two applications to a Philips 21' television set," Proceedings of the 1996 IEEE International Symposium on Electronics and the Environment, pp: 224–229.

Brennan, L., Gupta, S.M. and Taleb, K.N. (1994), "Operations planning issues in an assembly/disassembly environment," *International Journal of Operations and Production Management*, Vol: 14, No: 9, pp: 57–67.

Brinkley, A., Ray, J., Kirby, I. and Wadehra, L. (1996). "Life cycle inventory of PVC: Manufacturing and fabrication processes," Proceedings of the 1996 IEEE International Symposium on Electronics and the Environment, pp: 94–101.

Chandra, N.-D. (1994). "The recovery problem in production design," *Journal of Engineering Design*, Vol: 5, No: 1, pp: 65–86.

Choi Athelstan, K.Y. (1994), "A screening method for life-cycle inventory analysis for industrial material," Master of Applied Science Thesis, University of Windsor, Ontario, Canada.

"Enhancing design for disassembly: A benchmark of DfD software tools," Proceedings of the ASME Design Engineering Technical Conferences and Computers in Engineering Conference. Irvine, CA, August 18–22, 1996, Paper no. 96-DETC/DFM-1271, 1996.

Feldmann, K., Stefan, T. and Meedt, O. (1999). "Innovative disassembly strategies based on flexible partial destructive tools," *Annual Reviews in Control*, Vol: 23, pp: 159–164.

Ferrer, G. (2001). "On the widget remanufacturing operation," *European Journal of Operational Research,* Vol: 135, pp: 373–393.

Gu, P. and Sosale, S. (1999). "Product modularization for life cycle engineering," *Robotics and Computer Integrated Manufacturing*, Vol: 15, pp: 387–401.

Gungor, A. and Gupta, S.M. (1997). "An evaluation methodology for disassembly process," *Computers and Industrial Engineering,* Vol: 33, Nos: 1–2, pp: 329–332.

Gungor, A. and Gupta, S.M. (1998). "Disassembly sequence planning for products with defective parts in product recovery," *Computers and Industrial Engineering,* Vol: 35. Nos: 1–2, pp: 161–164.

Gungor, A. and Gupta, S.M. (1999). "Issues in environmentally conscious manufacturing and product recovery: A survey," *Computers and Industrial Engineering,* Vol: 36, pp: 811–853.

Hundal, M.S. (1996). "DFE: Current status and challenges for the future," ASME Design for Manufacturability Conference, Chicago, IL, pp: 89–98.

Kerr, W. and Ryan, C. (2001). "Eco-efficiency gains from remanufacturing: A case study of photocopiers remanufacturing at Fuji Xerox Australia," *Journal of Cleaner Production,* Vol: 9, pp: 75–81.

Krikke, H.R.A., Van Harten, P. and Schuur, C. (1998). "On a medium term recovery and disposal strategy for durable assembly products," *International Journal of Production Research,* Vol: 36. No: 1, pp: 111–139.

Krikke, H.R.A., Van Harten, P. and Shuur, C. (1999). "Busienss case Roteb: Recovery strategy for monitors," *Computers and Industrial Engineering,* Vol: 36, pp: 739–757.

Kriwet, A. Zussman, E. and Seliger, G. (1995). "Systematic integration of design-for-recycling into product design," *International Journal of Production Economics,* Vol: 38, pp: 15–22.

Kroll, E. and Carver, B. (1999). "Disassembly analysis through time estimation and other metrics," *Robotics and Computer Integrated Manufacturing,* Vol: 15, pp: 191–200.

Kroll, A. and Hanft, T. (1999). "Quantitative evaluation of product disassembly for recycling," *Research in Engineering Design,* Vol: 10, pp: 1–14.

Kroll, E., Beardsley, B. and Parulian, A. (1999). "A methodology to evaluate ease of disassembly for product recycling," *IIE Transactions,* Vol: 28, No: 10, pp: 837–845.

Kuo, T.C. (2000). "Disassembly sequence and cost analysis for electromechanical products," *Robotics and Computer Integrated Manufacturing,* Vol: 16, pp: 43–54.

Kuo, T.C., Huang, S.H., Zhang, H-C. (2001). "Design for manufacture and design for 'X': Concepts, applications and perspectives," *Computers and Industrial Engineering,* Vol: 41, pp: 241–260.

Lambert, A.J.D. (1999). "Linear programming in disassembly/clustering sequence generation," *Computers and Industrial Engineering,* Vol: 36, pp: 723–738.

Lee, S.G., Lye, S.W. and Khoo, M.K. (2001). "A multi-objective methodology for evaluating product end-of-life options and disassembly," *The International Journal of Advanced Manufacturing Technology,* Vol: 18, pp: 148–156.

Lucacher, R. (1996). "Competitive advantage and the environment: Building a framework or achieving environmental advantage," IEEE International Symposium on Electronics and the Environment, Dallas, TX.

Lund, R. (1983). "Remanufacturing, United States experience and implications for developing nations," MIT Center for Policy Alternatives Report, CPA/ 83-17, Cambridge, MA.

Mason, S. (1990). "Improving plant and machinery maintainability," *Applied Ergonomics,* Vol: 21, No: 1, pp: 15–24.

Maynard, H.B., Stegmerten, G.J. and Schwab, J.L. (1948). *Methods—Time Measurement.* McGraw Hill.

Mok, H.S., Kim, H.J. and Moon, K.S. (1997). "Disassemblability of mechanical parts in automobiles for recycling," *Computers and Industrial Engineering,* Vol: 33, Nos: 3–4, pp: 621–624.

Nagel, C. (1996). "Design for environment—A European logistics perspective," Green Electronics Green Bottom Line, *Environmentally Responsible Engineering,* pp: 215–223.

Nagel, C. and Meyer, P. (1999). "Caught between ecology and economy: End-of-life aspects of environmentally conscious manufacturing," *Computers and Industrial Engineering*, Vol: 36, pp: 781–792.

Ong, N.S. and Wong, Y.C. (1999). "Automatic subassembly detection from a product model for disassembly sequence generation," *The International Journal of Advanced Manufacturing Technology*, Vol: 15, pp: 425–431.

Ottman, J. (1998). *Green Marketing: Opportunity for Innovation.* NTC Business Books. Chicago.

Penev, K.D. and Ron, A.J. (1994). "Determination of a disassembly strategy," *International Journal of Production Research*, Vol: 34, No: 2, pp: 495–506.

Pollock, D. and Coulon, R. (1996). "Life cycle assessment of an inkjet print cartridge," Proceedings of the 1996 IEEE International Symposium on Electronics and the Environment, pp: 154–160.

Ron, Ad de and Penev, K. (1995). "Disassembly and recycling of electronic consumer products: An overview," *Technovation*, Vol: 15, No: 6, pp: 363–374.

Ross, V. and Choong, H. (1996). "HP deskjet printers with recycled plastics: A success story," Proceedings of the 1996 IEEE International Symposium on Electronics and the Environment, pp: 53–56.

Shu, L.H. and Flowers, W.C. (1999). "Application of a design-for-remanufacture framework to the selection of product life-cycle fastening and joining methods," *Robotics and Computer Integrated Manufacturing*, Vol: 15, pp: 179–190.

Shyamsundar, N. and Gadh, R. (1999). "Geometric abstractions to support disassembly analysis," *IIE Transactions,* Vol: 31, No: 10, pp: 935–946.

Society of Automotive Engineers (SAE) (1976). "Engineering design serviceability guidelines—Construction and industrial machinery."

Spicer, A. and Wang, M.H. (1997). "Environmental design industrial template (EDIT): A software tool for analysis of product retirement," *Journal of Cleaner Production*, Vol: 5, No: 3, pp: 193–198.

Srinivasan, H. and Gadh, R. (1998). "A geometric algorithm for single selective disassembly using the wave propagation abstraction," *Computer-Aided Design*, Vol: 30, No: 8, pp: 603–613.

Srinivasan, H. and Gadh, R. (2002). "A non-interfering selective disassembly sequence for components with geometric constraints," *IIE Transactions*, Vol: 34, No: 4, pp: 349–361.

Srinivasan, H., Figueroa, R. and Gadh, R. (1999). "Selective disassembly for virtual prototyping as applied to de-manufacturing," *Robotics and Computer- Integrated Manufacturing,* Vol: 15, No: 3, pp: 231–245.

Srinivasan, H., Shyamsundar, N. and Gadh, R. (1997), "A framework for virtual disassembly analysis," *Journal of Intelligent Manufacturing*, Vol: 8, No: 4, pp: 277–295.

Stevels, A.L.N (2000). "Green marketing of consumer electronics," *Electronics Goes Green*. Berlin, Germany.

Suga, T.K., Saneshige, J. and Fujimoto, J. (1996), "Quantitative disassembly evaluation," Proceedings of the 1996 IEEE International Symposium on Electronics and the Environment, pp: 19–24.

Taleb, K.N. and Gupta, S.M. (1997). "Disassembly of multiple product structures," *Computer and Industrial Engineering*, Vol: 32, No: 4, pp: 949–961.

Thierry, M., Salomon, M., Van Nunen, J., and Van Wassenhowe, L. (1995). "Strategic issues in product recovery management," *California Management Review*, Vol: 37, No: 2, pp: 114–135.

Trumble, W. "Selection of plastics using cost, structural and environmental factors," Green Electronics Green Bottom Line. *Environmentally Responsible Engineering*, pp: 137–142.

Ulrich, K. (1995). "The role of product architecture in the manufacturing firm," *Research Policy*, Vol: 24, pp: 419–440.

Veerakamolmal, P. and Gupta, S.M. (1998). "High-mix/low volume batch of electronic equipment assembly," *Computers and Industrial Engineering,* Vol: 35, No: 1–2, pp: 65–68.

Zhang, H.C., Tsai Kuo, C., Huitian, Lu, Samuel, H.H. (1997). "Environmentally conscious design and manufacturing: A state-of-the-art survey," *Journal of Manufacturing Systems*, Vol: 16. No: 5, pp: 352–371.

Zussman, E., Kriwet, A., Seliger, G. (1994). "Disassembly-oriented assessment methodology to support design for recycling," *Annals of the CIRP*, Vol: 43, No: 1, pp: 9–14.

4 Designing for Assembly

4.1 DEFINITION AND IMPORTANCE OF THE ASSEMBLY PROCESS

In general, a mechanical product comprises several constituent parts. Final products are an assembly of these constituents. If the product is sufficiently complex, it can be an assembly of several subassemblies. For instance, a complex product such as a television set or an automobile can be construed as being an assembly of different subassemblies. Each subassembly in this case consists of constituent parts that are functionally important.

Production rate is a direct function of the rate at which a product can be assembled. A product that is easy to assemble is likely to take less time to be put together and vice versa. This is the main idea behind designing products for assembly. From the point of view of sustainable product design, products are assembled as various stages of the product lifecycle. Initially, products are assembled when they are brand new. Later, they are reassembled after they have been repaired/refurbished or remanufactured. Thus, design for assembly (DfA) plays a critical role in the product life cycle.

The ease with which a product can be assembled is a function of design parameters that are both intensive and extensive in nature. Material properties constitute examples of intensive design parameters whereas physical attributes constitute examples of extensive design parameters. Design parameters include aspects of design such as shape, size, material compatibility, flexibility, thermal conductivity and so on. Assembly cycle time undergoes a significant reduction when a product has been designed with ease of assembly in mind. This is important from the point of view of sustainable product design since it leads to savings in material resources as well as human resources. This is such a critical issue that the design for manufacturing (DfM) philosophy has been extended to include assembly as one of the necessary design parameters. Hence the term: Design for manufacturing and assembly (DfMA).

The previous discussion has clearly emphasized the need for designing products for ease of assembly. In order to function as intended, it is necessary that the individual components and subassemblies of the product be fully functional. Also, it is important that not only should they function independently but also function as a single unit after the final product has been assembled. Functionality of a product is directly attributable to the spatial alignment of its various components as well as assembly tolerances. Thus, it is imperative that each individual component be designed to align and mate with its neighbor efficiently. This entails the design and processing of the component in a specific manner (with respect to shape, size, tolerances and surface finish). A product that has been designed for ease of assembly can be put together in less time. The subsequent cost of assembly is low as well.

4.2 DEFINITION OF DESIGN FOR ASSEMBLY

The main objective of DfA is to reduce the cost of the product assembly process. When DfA principles are applied to a product design, improved quality and reliability and a reduction in production equipment and part inventory is often the end result. These secondary benefits are so significant that they often end up overshadowing reduced assembly costs.

During the DfA process, the entire product design is analyzed from the point of view of the ease with which it can be put together. DfA may be defined as a process for improving product design for easy and low-cost assembly, which is achieved by means of concurrent focus on the dual aspects of functionality as well as assemblability. During the process of manufacturing a product, the following factors are the most important:

- Rate of production desired.
- Labor availability.
- Number of years the product is expected to last. In other words, it is important to take into consideration what is the product's market life.
- It is essential to carefully study the product's design from the point of view of functionality, manufacturability, ease of assembly and so on.

The practice of DfA as a distinct feature of design has only been relatively recently developed formally. However, informally, the practice of DfA has been around for a long time. It has been informally practiced since at least the 1960s. For instance, an internal manufacturing producibility handbook was published by General Electric (GE) in the 1960s. This handbook presents a set of guidelines and manufacturing data that designers can follow. These guidelines constituted many of the DfA principles. That particular terminology was never formally used. It was also not clearly distinguished from the remainder of the product development process.

4.3 TYPES OF ASSEMBLY METHODS

There are three types of assembly methods. Specifically, they can be characterized as manual assembly, automatic assembly and robotic assembly. Characteristics of each type of assembly method are described in the following section.

- **Manual assembly**: Manual Assembly is an assembly process in which operations are performed manually. This is usually accomplished with or without the aid of simple, general-purpose tools such as screw drivers and pliers. In this type of assembly process, the cost per unit is constant and the process required little initial investment. Thus, fixed costs are limited to investment in tools and assembly fixtures.

The process of manual assembly involves transferring parts to workbenches. This is typically where individual components are assembled into final products. In order to enable the assembly process and aid the worker, hand tools are commonly used. This is the most flexible and adaptable of assembly methods. But there is usually an upper limit to the production volume. Labor costs (including benefits, cases of workers compensation due to fatigue and injury, overhead for maintaining a clean, healthy environment, etc.) are higher in this type of assembly process.

- **Automatic assembly**: Automatic assembly is also referred to as fixed automation. It typically uses either synchronous indexing machines and part feeders or nonsynchronous machines. Parts are handled by a free transfer device. The system is generally built for a single product and the cost/unit decreases with increasing volume of production.

  Fixed or hard automation characteristically involves a custom-built machine that assembles one and only one specific product. This type of equipment entails a large amount of capital investment. As production volume increases, the fraction of the capital investment compared to the total manufacturing cost decreases. Indexing tables, parts feeders and automatic controls typify this inherently rigid assembly method. In some instances, automatic assembly is also referred to as "Detroit-type" assembly.
- **Robotic assembly**: Robotic assembly is best suited for those products whose production volume lies between manual and automatic assembly methods. This method of product assembly is the most flexible and can achieve volumes closer to the automatic assembly methods. Soft automation or robotic assembly incorporates the use of robotic assembly systems. This can take the form of a single robot, or a multistation robotic assembly cell with all activities simultaneously controlled and coordinated by a programmable logic controller (PLC) or computer. Although this type of assembly method can also have large capital costs, its flexibility often helps offset the expense across many different products.

  Robotic assembly offers numerous advantages over the competition (manual assembly and automatic assembly). For instance, it is flexible to changes in product design. Thus, if the product design changes, so can the assembly process. This is because the robot can be reprogrammed to account for changes in product design.

  Also, the production volume that can be handled by a robotic system is much higher than a high-speed automatic assembly machine. This is because a robot system can operate at much higher station cycle times than its fully automatic counterpart.

  A robotic system is also capable of handling various styles of the same product. For instance, a robot can be programmed to select only particular parts for the final assembly.

  Feeder systems are often used in high-speed automatic assembly. A feeder jam that is caused by a defective part in such systems contributes to greater loss than in the case of a robot with a longer cycle time. Robots can also be

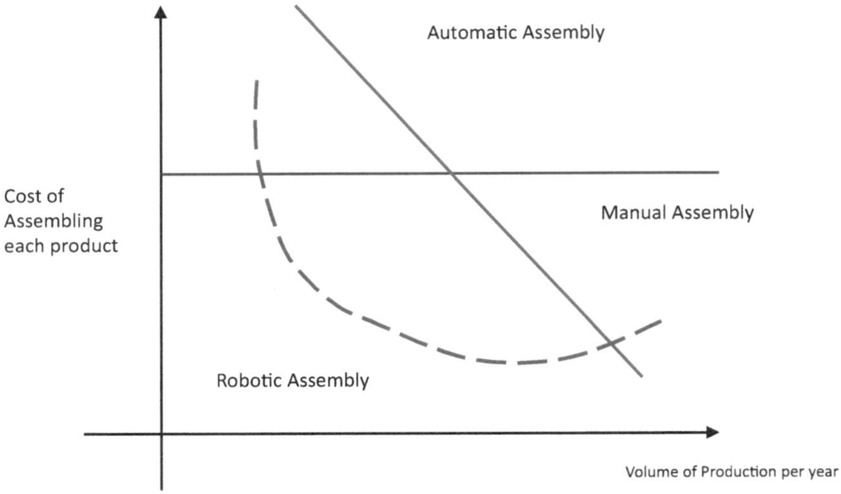

FIGURE 4.1 Comparing different methods of assembly based on type, production volume and costs involved.

programmed to anticipate problems by sensing them as they occur during the assembly process and then try the insertion process again.

One of the primary advantages of robotic assembly over automatic assembly lies in the fact that in the case of robotic assembly, parts can be presented in the form of patterns or arrays on part trays. Thus, the restrictions on part size that are inherent in the case of automatic assembly are done away with.

Figure 4.1 draws a comparison between the relative costs of the different methods of assembly by type as well as production volume.

The production range for each type of assembly is illustrated in Figure 4.2.

4.4 DIFFERENT MODES OF ASSEMBLY AND DESIGN GUIDELINES TO FACILITATE THEM

4.4.1 Design Guidelines for Manual Assembly

Several design guidelines have been developed to enable product design for ease of assembly. Specific guidelines have been developed for each type of assembly (such as manual assembly, automatic assembly and robotic assembly). This section presents some design guidelines for manual assembly.

- The need for decision making by the worker making the final adjustments should be eliminated if possible or it should at the very least be minimized. This reduces subjectivity, thereby enhancing the quality of the finished product.
- It should be ensured that the product can be easily accessed and is clearly visible.

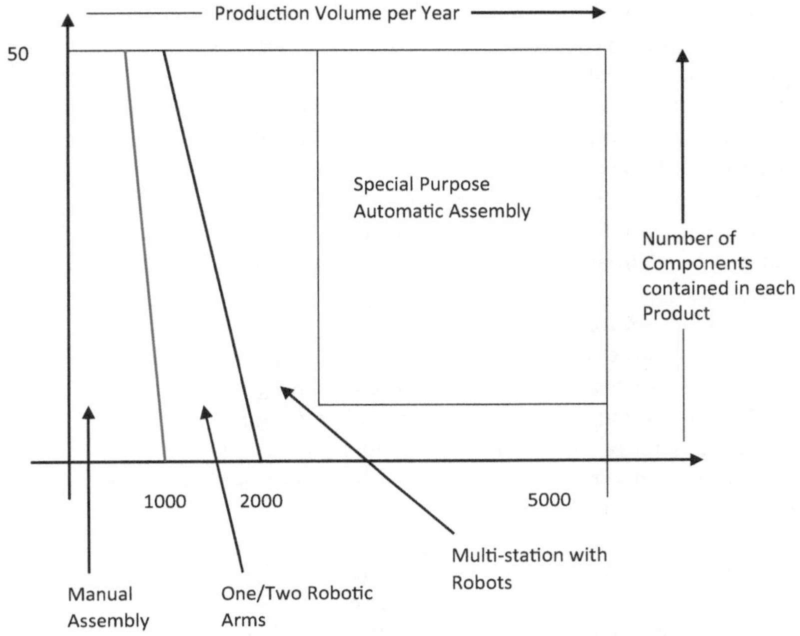

FIGURE 4.2 Relationship between different assembly methods and ranges of production.

- The need for assembly tools and special gauges can be eliminated by making the components self-aligning. The need for separate fasteners can be eliminated by incorporating design features such as snap fits. This results in a substantially faster and more economic assembly process.
- Multipurpose components can be used in order to minimize the total number of individual parts. This should be done only so long as the product functionality is not compromised.
- The number of individual parts can also be minimized by getting rid of excess parts. Two or more parts can also be combined into one if the functionality is not impaired.
- It is always important to minimize and avoid, if possible, the need to reorient the part while the assembly process is being performed. The process of inserting a part should always be simple. The need for rotating a part, releasing and regripping should be avoided as well. Vertical insertion takes advantage of gravity and that is always better than insertion along any other direction.

During the process of manual assembly, operators typically handle parts extensively. The process can be accomplished faster and more accurately if parts are designed in such a way as to avoid the need for such extensive handling. Manual handling of parts can be minimized by adhering to the following design guidelines.

- If possible, parts should be designed so that they have end-to-end symmetry. Parts can also have rotational symmetry about the axis of rotation. The design should always try and incorporate symmetry to the greatest extent possible.
- When it is not possible to design symmetric parts, obvious asymmetric features should be incorporated to facilitate ease of orientation and insertion.
- A part should be designed so it is secured as soon as possible post insertion. Holding down a part manually is not recommended.
- A part should be designed such that it is easily located prior to release. If the part is released before it has been located, it can pose problems with the assembly.
- Assembly costs for different types of commonly used fasteners are arranged in ascending order as follows: Snap fit, plastic bending, riveting and screwing. This factor is obviously important when securing and fastening methods for holding two or more parts together are being designed.
- From the point of view of parts insertion, the design should offer little to no resistance to the process of insertion. Chamfers can be provided to guide and facilitate insertion of two mating parts.
- If possible, parts should be assembled in a pyramidal fashion. Progressive assembly can be accomplished along one axis of reference. Assembling parts one on top of other is generally preferable.
- Design features that prevent jamming and entanglement of parts should be used as much as possible.

4.4.2 Design Guidelines for Automatic Assembly

The process of automatic assembly is significantly different from that of manual assembly. We have already seen this in the preceding section of this chapter. Owing to the obvious difference in the two processes, the guidelines for automatic assembly are quite different from those of manual assembly. The guidelines for automatic assembly are presented in this section.

- Unlike manual assembly, automatic assembly does not allow for manual manipulation of parts to perform assembly. For this reason, it is important to include self-aligning and self-locating features into the design. Design features such as chamfers, guide pins, dimples and cone and oval screws can be used for this purpose.
- The largest and most rigid part of the assembly should be used as a base or fixture. Other parts are stack assembled vertically on top of this base so that gravity can aid in the assembly process. A product thus assembled eliminates the need of using an assembly fixture. The assembly process is best performed one layer at a time. Another option is to divide the final assembly into several subassemblies which can then be plugged in later.
- In keeping with other Design for "X" (DfX) principles, a high percentage of standard parts should be used. Examples of standard components include fasteners such as bolts, screws, washers as well as standard modules and subassemblies.

- The process of parts nesting, shingling or tangling can complicate and delay the assembly process. Any design features that lead to such behavior should be avoided. This is similar to manual assembly.
- Reorienting assemblies may require a separate workstation or machine. This increases costs and should be avoided.
- Automatic assembly can be facilitated by designing parts, so they are presented or admitted to the assembly machine in the appropriate orientation after the minimum possible time spent in the feeder.
- Symmetrical or clearly asymmetrical parts can be easily oriented.
- Parts with a low center of gravity (CoG) can be fed easily.
- Flexible, fragile and abrasive parts should be avoided. It should be ensured that parts are sufficiently strong and rigid to enable them to withstand the forces experienced during feeding, assembly and use.

4.4.3 Design Guidelines for Robotic Assembly

- Part type and part count should be reduced as much as possible.
- In view of the poor repeatability of robot manipulators, features such as lips, leads, chamfers and so on should be carefully designed.
- Self-locating parts should be used during assembly if it is known that they will not be secured immediately upon insertion.
- The same robot gripper should be able to grip and insert all components. Part design should facilitate this. Process efficiency is compromised if grippers and tools need to be changed frequently.
- Product design should lend itself to vertical assembly directly from above.
- As with manual and automatic assembly, there should be minimal need for part reorientation. Part design should enable easy gripping by the robot arm.
- Individual components should be designed so as to promote ease of handling from bulk. Parts that nest, tangle, are flexible, thin, heavy, exceedingly large or small and parts that are slippery should be avoided.

Some of the principles that are used to facilitate ease of manual assembly are depicted pictorially in Figures 4.3–4.7.

The principles of high-speed automatic assembly and robotic assembly are depicted pictorially in Figures 4.8–4.13.

The next section presents an in-depth discussion of three DfA methods, namely:

- The Hitachi assembly evaluation method.
- The Lucas DfA method.
- The Boothroyd Dewhurst DfA method.

4.5 METHODOLOGIES FOR EVALUATING DFA

This section provides an overview of three distinct methods for designing products for ease of assembly. These methods enable evaluating the efficiency of a product design from the perspective of product assembly. Comparison of two different

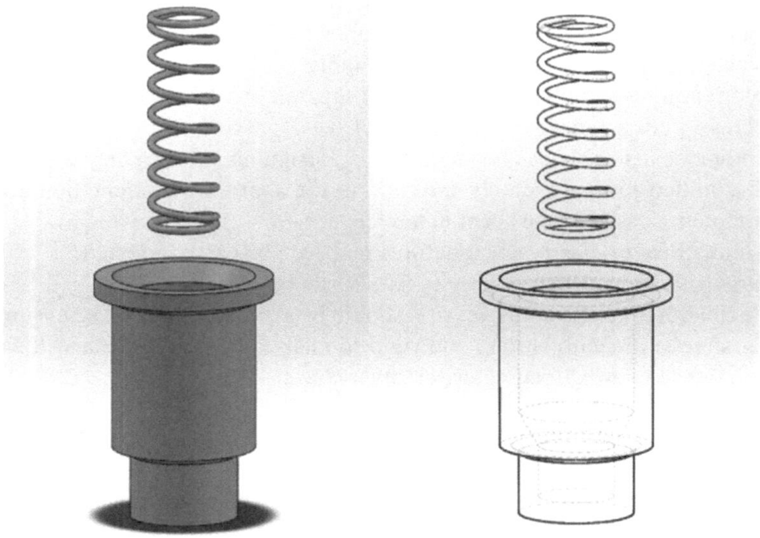

FIGURE 4.3 Parts falls into place due to provision of chamfer (Adapted from Boothroyd, 2005).

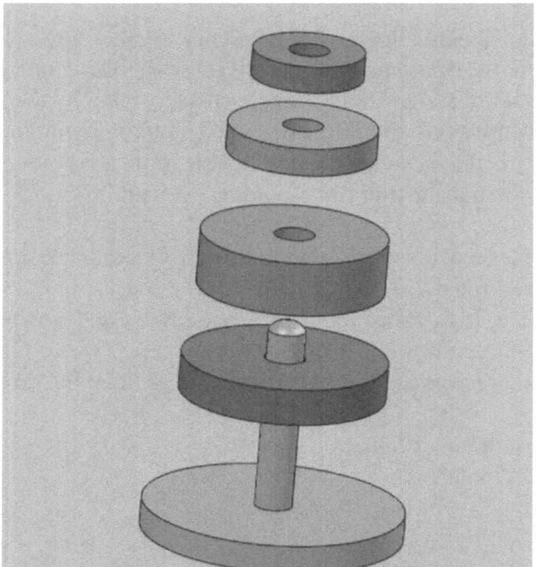

FIGURE 4.4 Assembly around a single axis (Adapted from Boothroyd, 2005).

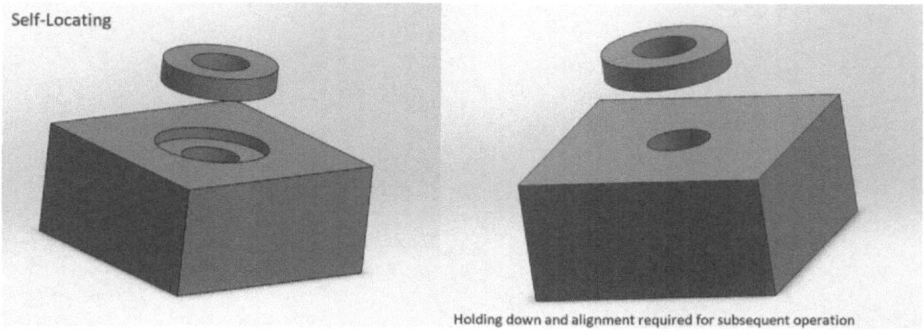

FIGURE 4.5 Self-locating features avoid need for holding down and alignment (Adapted from Boothroyd, 2005).

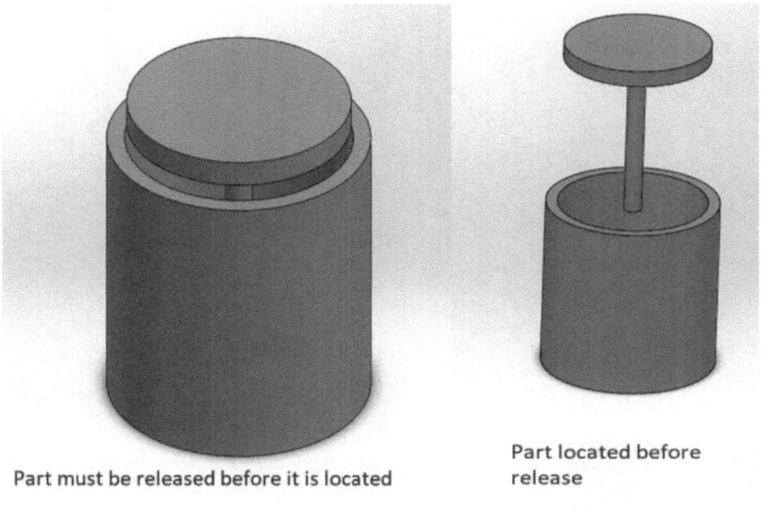

FIGURE 4.6 Part design for ease of insertion (Adapted from Boothroyd, 2005).

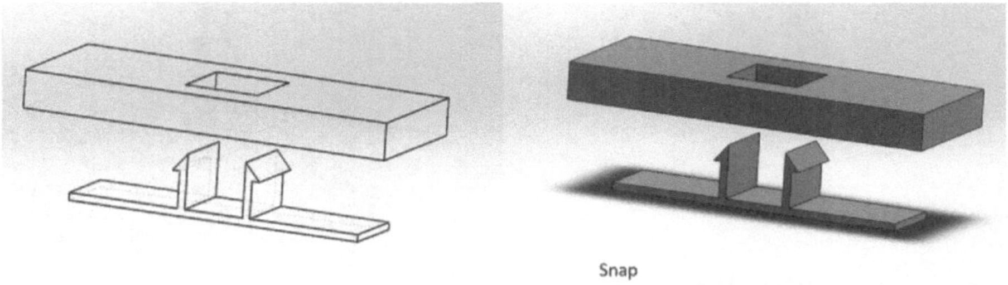

FIGURE 4.7A Snap fits are the "least expensive" form of assembly (Adapted from Boothroyd, 2005).

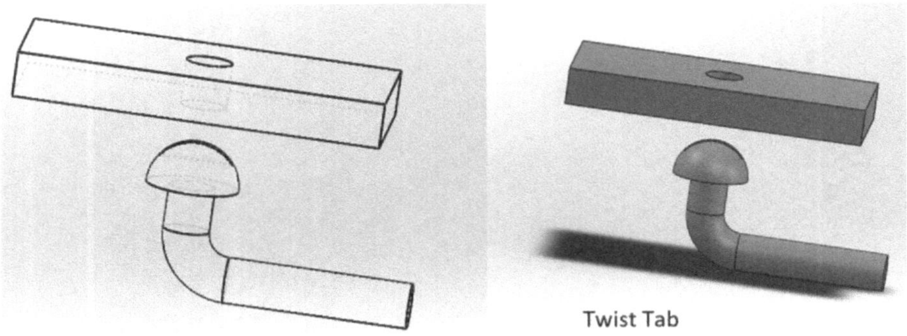

FIGURE 4.7B Twist tabs are more expensive than snap fits but less expensive than rivets (Adapted from Boothroyd, 2005).

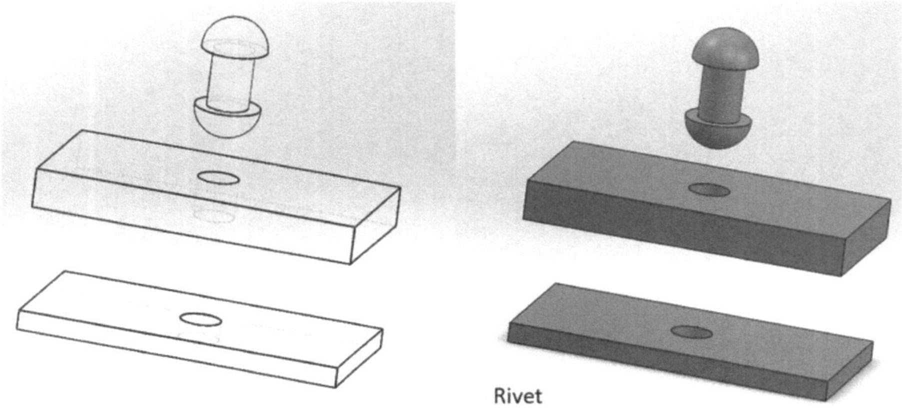

FIGURE 4.7C Rivets are more expensive than twist tabs but less expensive than screws (Adapted from Boothroyd, 2005).

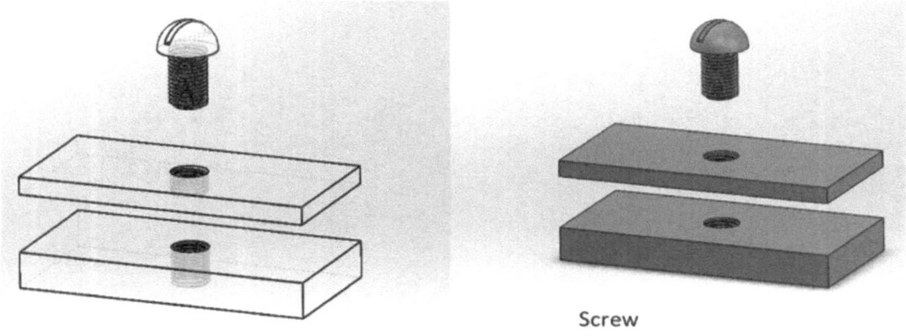

FIGURE 4.7D Screws are the "most expensive" form of fasteners (Adapted from Boothroyd, 2005).

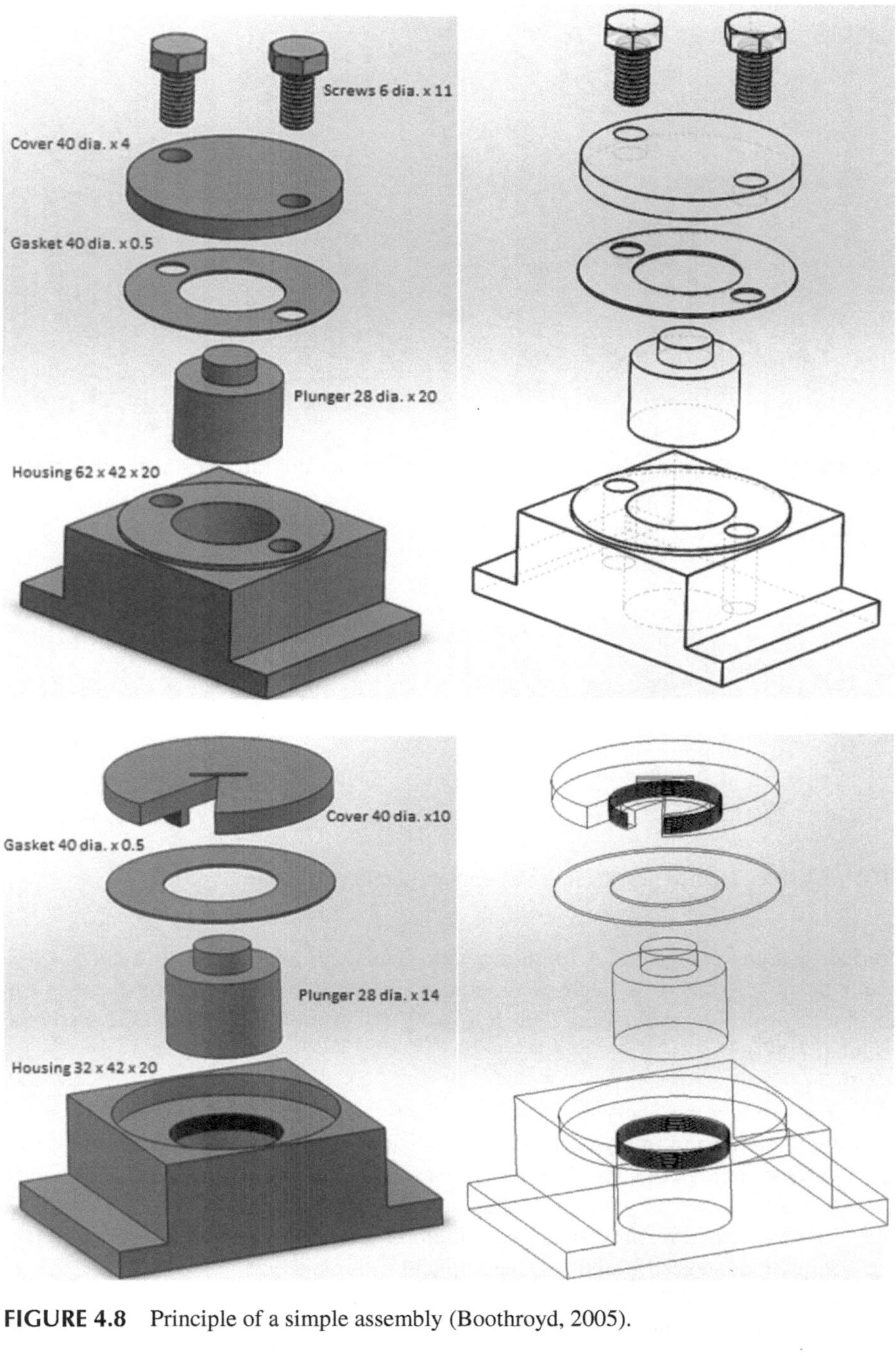

FIGURE 4.8 Principle of a simple assembly (Boothroyd, 2005).

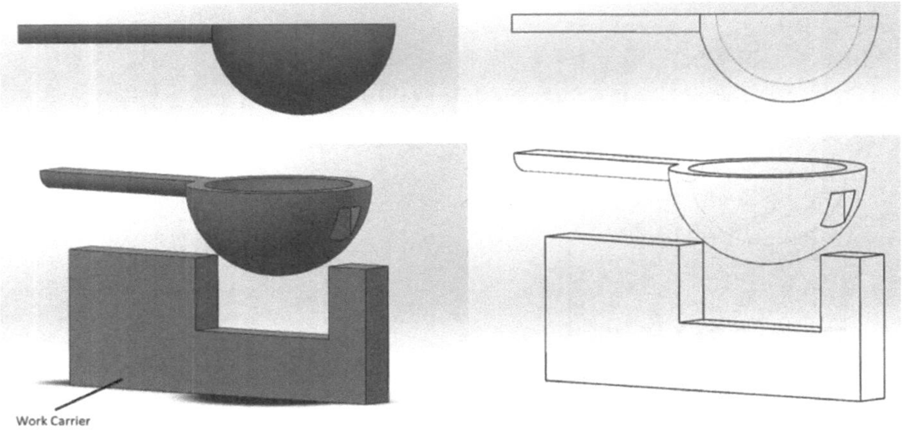

FIGURE 4.9 Base part design that enables easy mount on work carrier (Boothroyd, 2005).

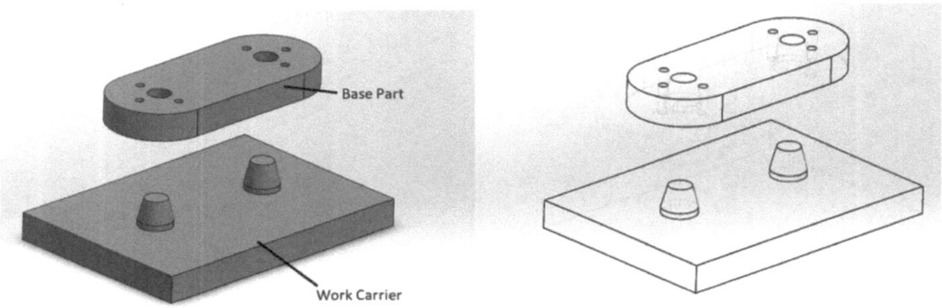

FIGURE 4.10 Tapered pegs facilitate assembly (Boothroyd, 2005).

product designs using one of these methodologies can enable one to ascertain the better design of the two. Each methodology presented in this section focuses on an objective design evaluation. The following are some of the aforesaid methods (Huang, 1996):

- *The Hitachi assembly evaluation method*: The objective of this method is to facilitate design improvements by identifying weaknesses in the design at the earliest stage in the design process by using an assemblability evaluation score (E) and an assembly cost ratio (K).
- *The Lucas DfA method*: The Lucas method seeks to analyze the design in three sequential stages: functional, feeding and fitting analyzes.
- *The Boothroyd Dewhurst DfA method*: This method seeks to reduce the number of parts by taking into account factors such as manual handling and manual insertion times.

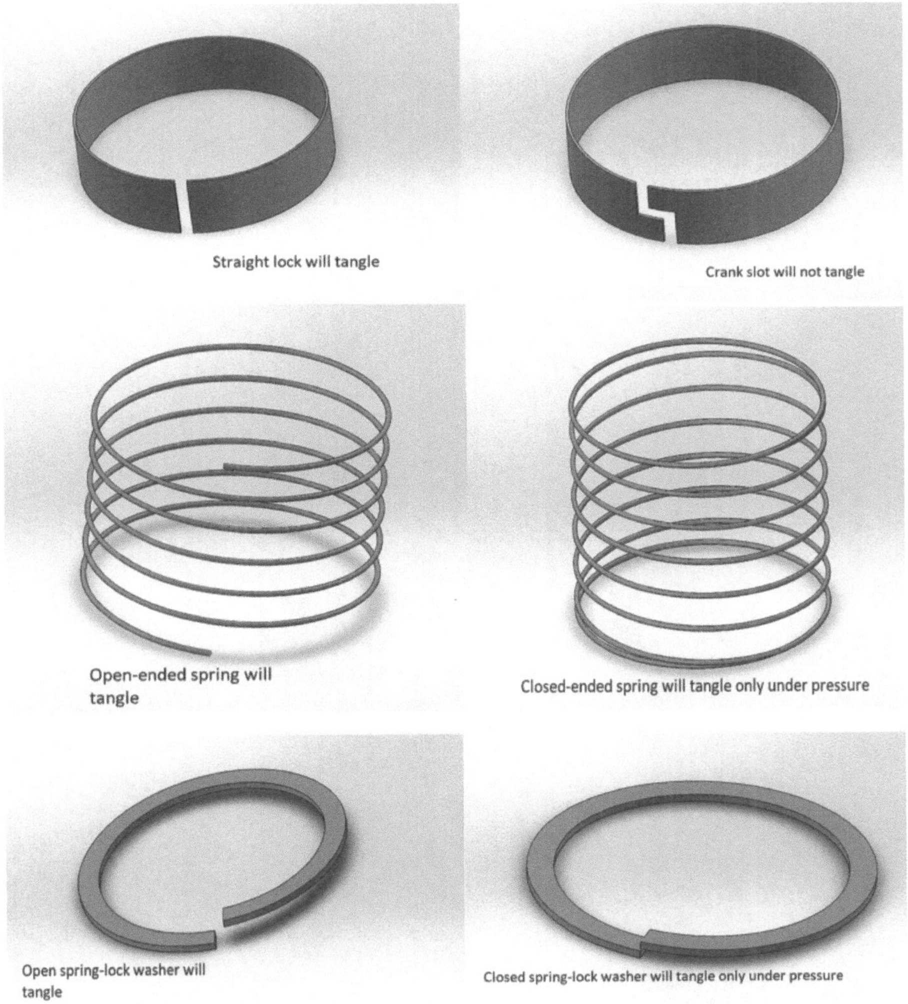

FIGURE 4.11 Part redesign to prevent tangling or nesting of parts (Boothroyd, 2005).

4.5.1 The Hitachi Assemblability Evaluation Method

The objective of this method is to facilitate design improvements by identifying weaknesses in product design at the earliest possible stage in the design process. It uses two primary indicators to achieve this objective:

- An assemblability evaluation score ration E, which is used to access design quality by determining the difficulty of operations.
- An assembly cost ratio K, used to project elements of assembly cost.

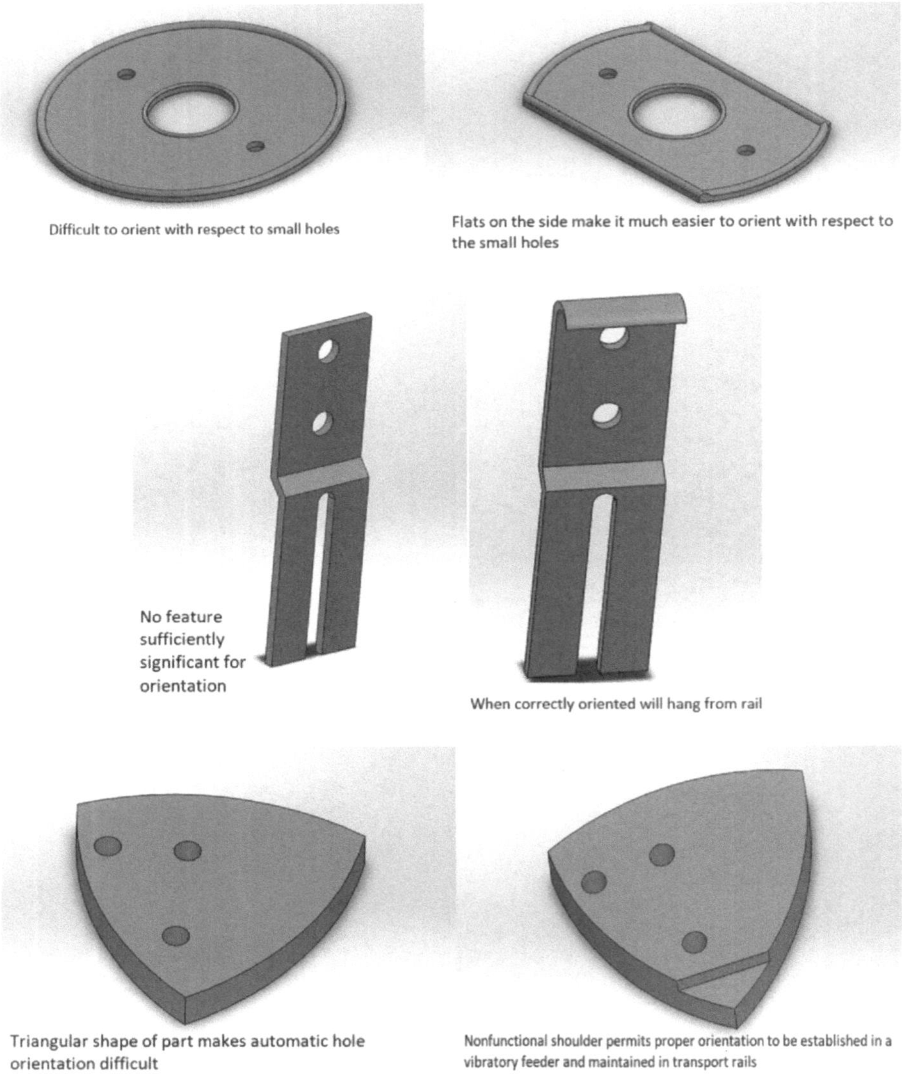

FIGURE 4.12 Asymmetrical features assist in part orientation (Boothroyd, 2005).

The Hitachi method considers both cost as well as quality. This implies that a low-cost design is not necessarily the lowest cost design. Alternatively, a good design may be too expensive. This is the only evaluation method that considers product design economics. For that reason, it is not purely technical in nature.

The following demonstrates the flow of logic in making design decisions using the Hitachi evaluation method.

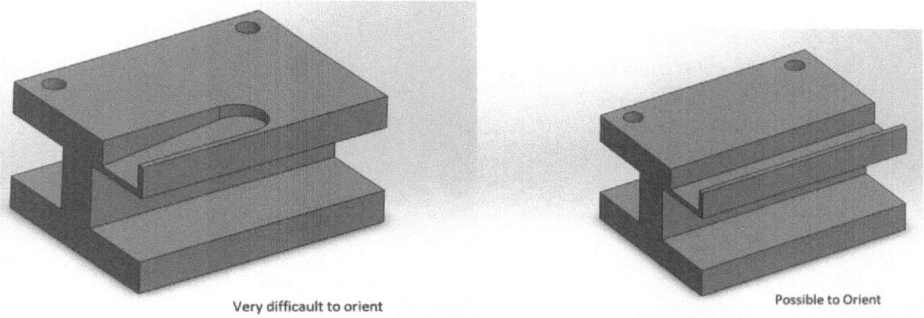

FIGURE 4.13 Part feeding, and orientation facilitated through a design change (Adapted from Boothroyd, 2005).

4.5.2 The Procedure for Design Analysis Using Hitachi Evaluation Method

The general procedure of design evaluation is presented stepwise as follows:

- All assembly operations are categorized into 20 elemental assembly tasks. Each task is assigned a symbol indicating the task content. Each task relates specifically to insertion and fastening processes and not to parts handling.
- Each of the elemental tasks is subject to a penalty score. This score reflects the degree of difficulty of the task. The penalty scores are obtained from analysis of shop floor data and are constantly revised to reflect changes in technology and methods. The penalty scores are then ranked, and all are compared to the elemental task with the lowest penalty score. For example, the simple task of placing an object on top of another object without requiring much accuracy is assigned a penalty score of 0 due to the inherent ease with which the task can be performed. Conversely, the more complicated task of soldering (assembly method) is assigned a much higher penalty score of 20.
- Factors which influence elemental tasks are extracted as coefficients and the penalty scores are modified accordingly.
- Attaching (contacting) conditions appropriate for each part are expressed using further Assembly Evaluation method (AEM) symbols.
- The total of the various penalty scores for an individual component are then modified by the contacting coefficients (as described) and subtracted from the best possible score (100) to give the assemblability evaluation score for the part.
- The total score for the product is defined as the sum of the assemblability scores for individual tasks divided by the total number of tasks.
- Typically, a score of 80 or above and a "K" value of 0.7 (implying savings of 30%) or lower is acceptable.

4.5.3 The Lucas DfA Evaluation Method

The Lucas DfA method was developed in the early 1980s by the Lucas Corp. in the United Kingdom. The Lucas method is based on a "point scale." The point scale

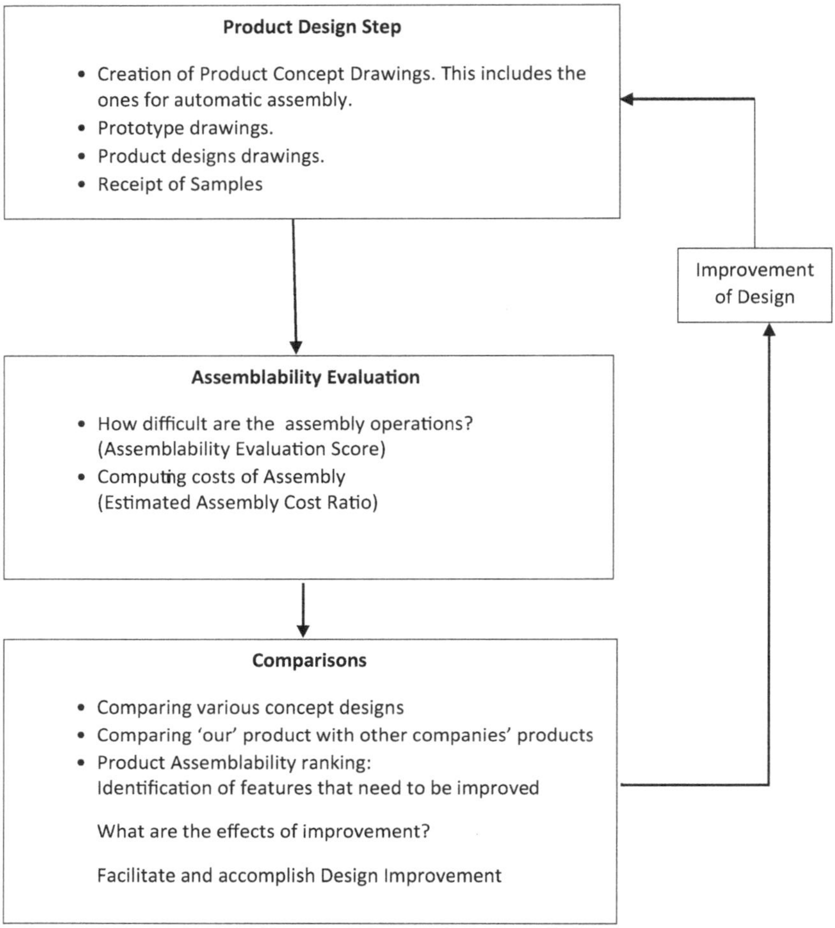

FIGURE 4.14 Assemblability evaluation and design improvement flow diagram for the Hitachi evaluation method.

assigns a relative measure of the difficulty to each assembly task. This method is based on three separate analyzes that are performed sequentially. These analyzes are described by means of the assembly sequence flowchart (ASF). Figure 4.15 illustrates the Lucas design for mechanical assembly (DfMA) procedure.

1 Design specification of the product.
2 Analysis of the product.
3 Functional analysis (this is the first Lucas analysis). If functional analysis presents problems, go back to step 2.
4 Feeding analysis (second Lucas analysis).
5 Fitting analysis (third Lucas analysis).

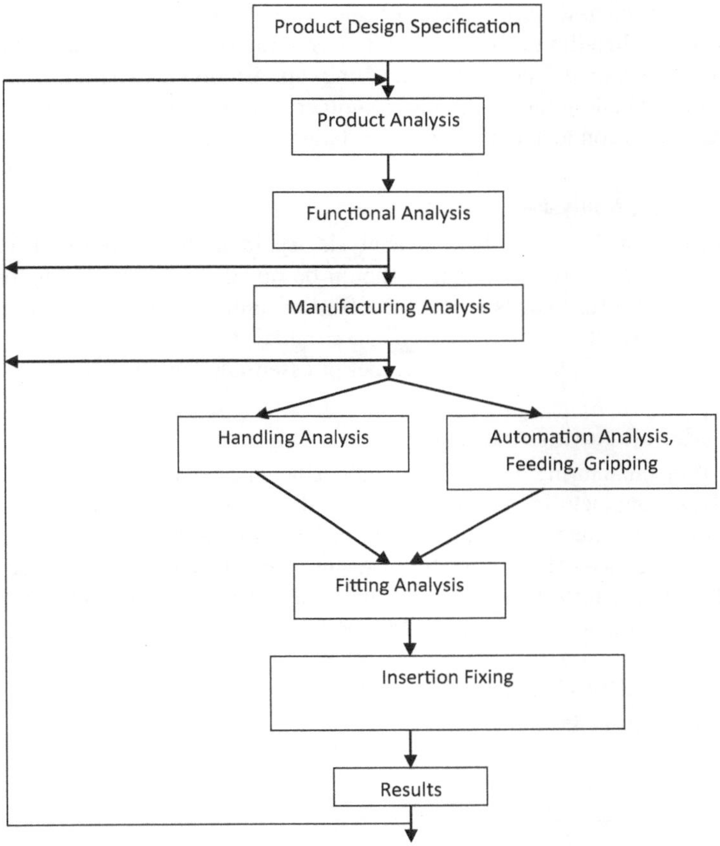

FIGURE 4.15 Assembly sequence flowchart for Lucas DfA evaluation model.

6 Assessment.
7 Possibly return to step 2 if problems have been identified in the analyzes.

4.5.3.1 Functional Analysis

During the first part of the evaluation, the product is analyzed functionally. Components are divided into two groups. The first group includes components that perform a primary function. Thus, they exist for a fundamental reason. These components are considered essential or "A" parts. The second group "B" components are nonessential components such as fasteners and locators.

The design efficiency can be computed using the formula: DE = A/(A+B) *100

The target efficiency should be at least 60%.

4.5.3.2 Feeding Analysis

Feeding analysis forms the second part of this evaluation system. This is concerned with the problems associated with handling components and subassemblies until they are admitted to the assembly system. By answering a group of questions regarding the size, weight, handling difficulties and orientation of a part, its feeding/handling index can be calculated. The feeding/handling ratio is computed as follows:

F/H ratio = (Feeding/handling index)/number of essential components

Fitting analysis is conducted similarly. The target value is 2.5

4.5.3.3 Fitting Analysis

The fitting analysis is like the feeding analysis. Again, a fitting index of 1.5 is a goal value for each assembly. However, it should be noted that there is usually greater variance in the fitting indices than in the feeding indices. Again, an overall fitting ratio of 2.5 is desired.

Fitting ratio = Total fitting index/number of essential components

4.5.3.4 Manufacturing Analysis

The cost of manufacturing each component is calculated in the last part of the Lucas method. This manufacturing cost can influence the choice of material and the process by which the part is made. Although not a true "costing" of the part, this method does help guide designers by giving a relative measure of manufacturing cost. Values of each of the following coefficients are derived from detailed tables developed for the purpose.

The part manufacturing cost index $M_i = R_c P_c + M_c$ where

R_c: $C_c C_{mp} C_s$ (C_t **or** C_f) is the relative cost
C_c: Complexity factor

TABLE 4.1
Lucas DfA method: Manual handling analysis

Handling index = A+B+C+D

A. Size and weight of part		**B. Handling difficulties**	0.4
		Delicate	
Very small—requires tools	1.5	Flexible	0.6
Convenient—hands only	1	Sticky	0.5
Large and/or heavy	1.5	Tangible	0.8
requires more than one hand		Severely nest	0.7
Large and/or heavy	3	Sharp/Abrasive	0.3
requires hoist or two people			
		Untouchable	0.5
		Gripping problem/slippery	0.2
		No handling difficulties	0
C. Orientation of part		**D. Rotational orientation of part**	
Symmetrical, no orientation required	0	Rotational symmetry	0
End to end, easy to see	0.1	Rotational orientation, easy to see	0.2
End to end, not visible	0.5	Rotational orientation, hard to see	0.4

TABLE 4.2
Lucas DfA method: Manual fitting analysis

A. Part placement and fastening	**Score**
Self-holding orientation	1.0
Requires holding	2.0
Plus, one of the following	
Self-securing (such as snaps)	1.3
Screwing	4.0
Riveting	4.0
B. Direction of process	
Straight line from above	0
Straight line not from above	0.1
Not a straight line	1.6
Bending	4.0
C. Insertion	
Single insertion	0
Multiple insertions	0.7
Simultaneous multiple insertions	1.2
D. Access and/or vision	
Direct	0
E. Alignment	
Easy to align	0
Difficult to align	0.7
F. Insertion force	
No resistance to insertion	0
Resistance to insertion	0.6
Restricted	1.5

C_{mp}: Material factor
C_s: Minimum section
C_t: Tolerance factor *or* C_f: Finish Factor (whichever is greater)
P_c: Processing cost
M_c: V C_{mt}, W_c is the material cost
V: Volume (mm^3)
C_{mt}: Material cost
W_c: Waste coefficient

4.5.4 The Boothroyd Dewhurst DfA Method

This method is based on two principles:

- The application of criteria to each part to determine if it should be separate from all other parts.

- The estimation of the handling and assembly costs for each part using the appropriate assembly process.

The Boothrooyd Dewhurst method relies on an existing design which is iteratively evaluated and improved. The process follows the following steps:

- Select an assembly method for each part.
- Analyze the parts for the given assembly methods.
- Refine the design in response to shortcomings identified by the analysis.
- Refer to step 2 until the analysis yields a satisfactory design.

The analysis is generally performed using a specific worksheet. Tables and charts are used to estimate the part handling and part insertion time. Each table is based on a two-digit code that is in turn based on a part's size, weight and geometric characteristics. Handling and insertion times are a function of the following component parameters. Each of these parameters directly affects the assembly process by simplifying or complicating it:

- What is the component size?
- What is the thickness of the component?
- What is the weight of the component?
- What is the degree to which the component tends to nest?
- What is the degree to which the component tends to tangle?
- How fragile is the component?
- How flexible is the component?
- How slippery is the component?
- How sticky is the component?
- Is it necessary to use two hands to perform assembly?
- Is it necessary to use specialized grasping tools to perform assembly?
- Is it necessary to optically magnify components to perform assembly?
- It is necessary to use mechanical assistance devices to perform assembly?

Nonassembly operations are also included in the worksheet. For example, extra time is allocated for each time the assembly is reoriented.

Next, all parts are evaluated on the basis of whether each part is really necessary in the assembly by asking the following questions:

- Does the part move relative to another part?
- Are the material properties of the part necessary?
- Does the part need to be a separate entity for the sake of assembly?

The list of all parts is then evaluated to obtain the minimum number of theoretically needed parts, denoted by N_m.

The list of all parts is then evaluated to obtain the minimum number of theoretically needed parts, denoted by N_m. In Table 4.3, in column "I," the number 1 is used to

TABLE 4.3
Boothroyd Dewhurst method to evaluate DfA (Boothroyd)

A	B	C	D	E	F	G	H	I	Name of assembly
Part ID	Number of consecutive identical operations	2-digit handling code	Manual handling time per part	2-digit insertion code	Manual insertion time per part	Operation time (BD + F)	Operation cost	Essential part?	
Totals:						T_m =	C_m =	N_m=	

represent that a part is essential, and 0 is used to represent that a part is not essential. The method then assumes that the ideal assembly time for a part is 3 seconds. Given that assumption, the design efficiency can be calculated as: $(3s \times N_m)/T_m$.

This DfA method utilizes a classification system consisting of two digits. Each digit can take on values from 0 to 9. The first digit can be divided into four groups as follows:

- Group I: If the first digit is 0–3, it indicates that the part is of a nominal size and weight and is easy to grasp and manipulate with one hand without using any tools.
- Group II: If the first digit is 4–7, it indicates parts that require use of grasping tools due to small size.
- Group III: If the first digit is 8, it is indicative of parts that severely nestle or tangle in bulk.
- Group IV: If the first digit is 9, it indicates parts that require two hands, two workers or require mechanical assistance in handling.

Groups I and II are subdivided into categories that represent the degree of orientation that is required. This is based on part symmetry.

As far as the second digit of the handling code is concerned, it is based on attributes such as flexibility, slipperiness, fragility and tendency to nest. It is also a function of the group divisions of the first digit. This is described as follows:

- Group I: If the first digit is 0–3, then the second digit will classify the thickness and size of the part.
- Group II: If the first digit is 4–7, then the second digit will classify part thickness, tool required for part handling and the necessity to optically magnify the handling process.
- Group III: If the first digit is 8, then the second digit will classify the part symmetry and size.
- Group IV: If the first digit is 9, then the second digit will classify the weight, symmetry and interlocking characteristics of parts in bulk.

When manual insertion and fastening methods are being considered, the Boothroyd and Dewhurst method recognizes that there are a limited number of basic assembly tasks such as peg in hole, weld, rivet, screw and so on. Manual insertion times are directly affected by the following design features:

- What is the accessibility of the assembly location?
- How easily can the operation tool be operated?
- How easily visible is the assembly location?
- How easily can the part be aligned and positioned during assembly?
- How deep should the part be inserted?

The manual insertion and fastening coding system is described as follows:

- Group I: If the first digit is 0–2, then the part is not secured immediately after insertion.
- Group II: If the first digit is 3–5, then the part secures itself or secures another part post insertion.
- If the first digit is 9, then the process is dealing with parts that are already in place.

Groups I and II are further divided into classes which take into account the effect of obstructed access and/or restricted visibility on assembly time. The second digit of the assembly code is based on group divisions of the first digit as follows:

- Group I: If the first digit is 0–2, then the second digit addresses the ease with which parts can be engage. It also considers whether it is essential to hold down the part to maintain orientation or location.
- Group II: If the first digit is 3–5, then the second digit classifies the ease with which parts can be engaged. This occurs during the fastening operation such as a snap fit, screwing or an assembly that involves a plastic deformation process.
- Group III: If the first digit is 9, then the second digit is used to classify mechanical, chemical and metallurgical processes.

An example of estimated times for the manual insertion process where a part is added but not secured is depicted in Tables 4.4 and 4.5.

One of the drawbacks of the Boothroyd and Dewhurst method is that it can be time consuming owing to the amount of intricate detail involved in the analysis procedure. However, a software package has been developed in order to accelerate the application of this process.

The Boothroyd Dewhurst method has been known to reduce the total number of individual components in an assembly. However, this has often been achieved at the cost of part complexity. Part complexity does not lend itself easily to manufacturability. Similarly, complex parts are inherently more difficult to disassemble as well.

In order to better understand the Boothroyd and Dewhurst method, consider the example of a controller assembly as depicted in Figure 4.16.

TABLE 4.4
Estimated time (in seconds) for manually inserting parts where neither the part nor any other part is finally secured immediately

		No holding down the part is required to maintain orientation and location				Necessary to hold down the part during subsequent processes to maintain orientation and location			
		Easy to align and position during the assembly process		Not easy to align or position during assembly		Easy to align and position during the assembly process		Not easy to align or position during assembly	
		No resistance to insertion	Resistance to insertion	No resistance to insertion	Resistance to insertion	No resistance to insertion	Resistance to insertion	No resistance to insertion	Resistance to insertion
		0	**1**	**2**	**3**	**6**	**7**	**8**	**9**
Part and assembly tool can reach the location easily	0	1.5	2.5	2.5	3.5	5.5	6.5	6.5	7.5
Part and assembly tool cannot reach the location due to obstructed access or restricted vision (2)	1	4	5	5	6	8	9	9	10
Part and assembly tool cannot reach the location due to obstructed access or restricted vision (2)	2	5.5	6.5	6.5	7.5	9.5	10.5	10.5	11.5

Source: Assembly Automation and Product Design (Boothroyd).

TABLE 4.5
Worksheet for controller assembly

Item	Number of items	Manual handling code	Handling time per item in seconds	Manual insertion code	Insertion time per item in seconds	Total operation time in seconds	Figures for min parts	Description
1. Pressure regulator	1	30	1.95	00	1.50	3.45	1	Secure in fixture
2. Metal frame	1	30	1.95	06	5.50	7.45	1	Add metal frame
3. Nut	1	00	1.13	39	8.00	9.13	0	Add and fasten screw
4. Reorientation	1			98	9.00	9.00		Reorient and adjust
5. Sensor	1	30	1.95	08	6.50	8.45	1	Add sensor
6. Strap	1	20	1.80	08	6.50	8.30	0	Add and hold down strap
7. Screw	2	11	1.80	39	8.00	19.60	0	Add and fasten screw
8. Application of tape	1			99	12.00	12.00		Apply tape
9. Adapter nut	1	10	1.50	49	10.50	12.00	0	Add and fasten screw
10. Tube assembly	1	91	3.00	10	4.00	7.00	0	Add and fasten screw
11. Fastening the screw	1			92	5.00	5.00		Fasten screw
12. PCB assembly	1	83	5.60	08	6.50	12.10	1	Add and hold down
13. Screw	2	11	1.80	39	8.00	19.60	0	Add and fasten screw
14. Connector	1	30	1.95	31	5.00	6.95	0	Add and snap fit connector
15. Earth lead	1	83	5.60	31	5.00	10.60	0	Add and snap fit earth lead
16. Reorientation of assembly	1			98	9.00	9.00		Reorient assembly and adjust
17. Knob assembly	1	30	1.95	08	6.50	8.45	1	Add and fasten screw
18. Fastening the screw	1			92	5.00	5.00		Fasten the screw
19. Plastic cover	1	30	1.95	08	6.50	8.45	0	Add and hold down plastic cover
20. Reorienting the assembly	1			98	9.00	9.00		Reorient and adjust assembly
21. Screw	3	11	1.80	49	10.50	36.90	0	Add and fasten screw

Source: Assembly Automation and Product Design (Boothroyd).

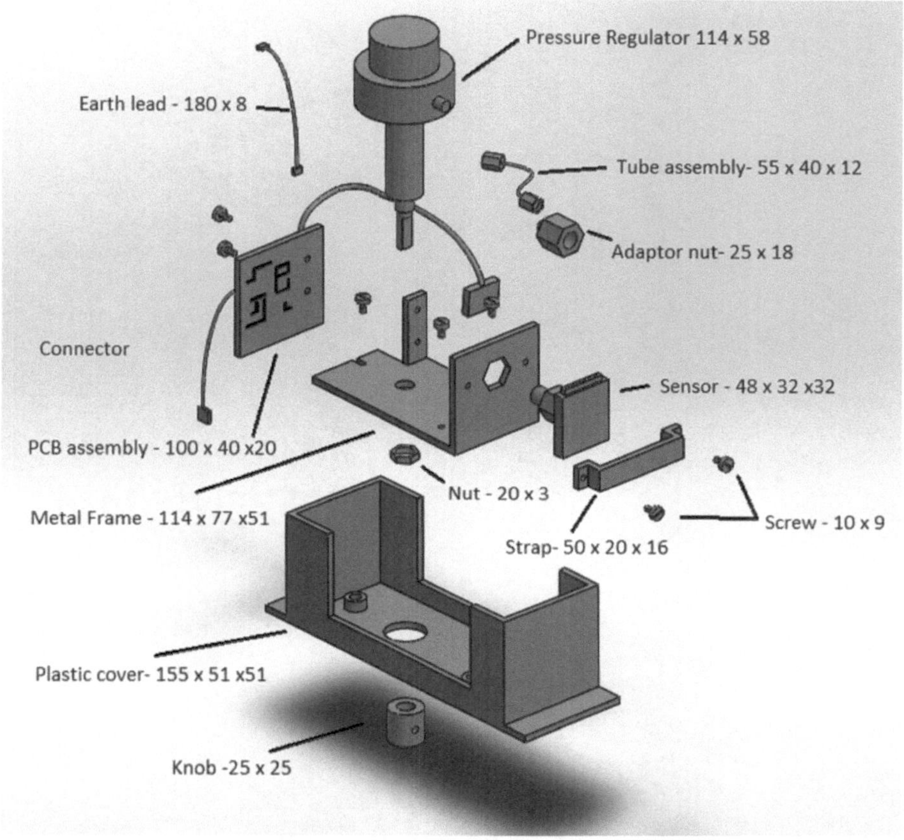

FIGURE 4.16 Schematic of controller assembly (Boothroyd, 2005).

The assembly process begins by placing the pressure regulator into a fixture and the metal frame is secured onto it using a securing nut. This assembly is then turned over so that other items can be added onto the metal frame. A step-wise analysis of the assembly process is presented in Table 4.5. It is to be noted that the total operation time is the sum of handling and insertion times multiplied by the number of items. Thus, it will be observed that the total operation time for the strap is 8.3 seconds.

The criteria for establishing the theoretical minimum part count are applied after the pressure regulator, metal frame, nut and sensor have already been assembled. Note that the strap does not move relative to the aforementioned parts. Thus, it could, in theory, be combined with any of them. Also, it is not necessary that the strap be made from a different material. Because the sensor has been purchased, it is not possible to change its design. Also, bear in mind that it is not necessary for the strap to be separate from the sensor to facilitate assembly of the sensor. Thus, the strap can potentially be eliminated. For this reason, 0 is placed in the column corresponding to the minimum number of parts.

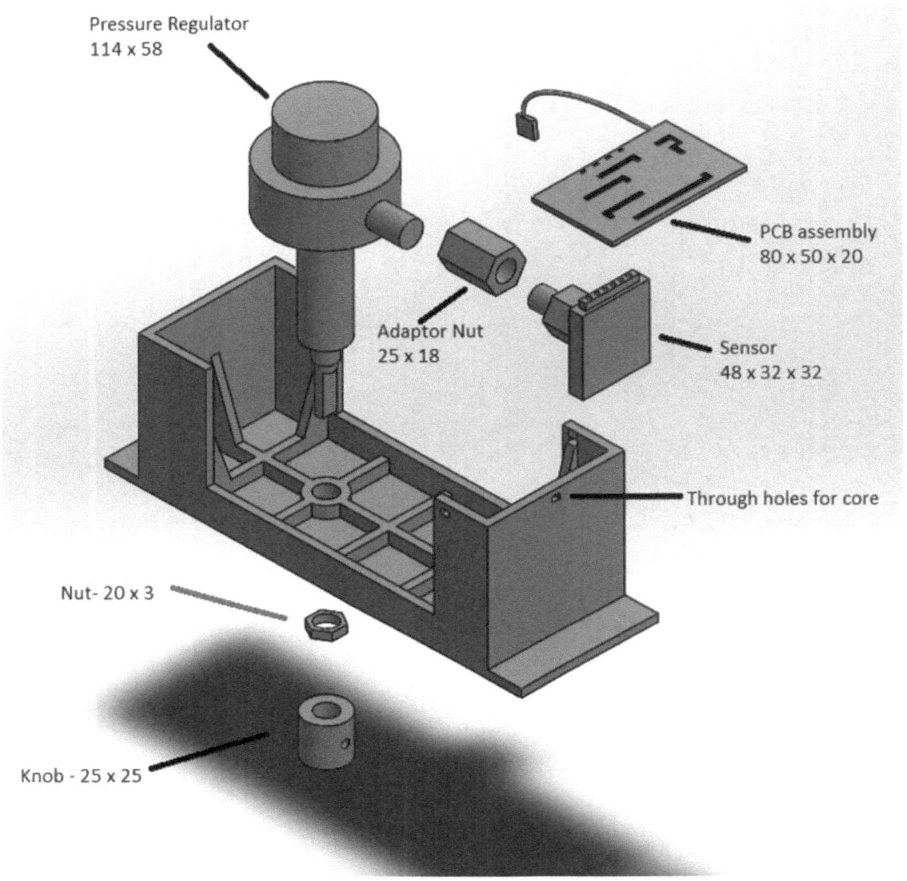

FIGURE 4.17 Controller assembly redesigned using Boothroyd DfA method (Boothroyd, 2005).

Using the principles of DfA, a number of possible design changes can be recommended for the controller assembly depicted in Figure 4.17. These design changes are presented in Table 4.6.

If adopted, the suggested design changes could result in saving at least 149.4 seconds of assembly time. This amounts to a 66% reduction of the total assemble time. Also, many hardware items could potentially be eliminated. This, in turn, leads to reduced parts cost. A schematic of the altered design is depicted in Figure 4.17.

As a consequence of applying the DfA methodology, the new assembly time for the modified controller assembly is now 84 seconds. The technical and economic consequences of the design changes such as impact on functionality and cost effectiveness should be considered at this stage. Documented savings in materials, manufacturing and assembly indicate direct costs. In order to compute total costs, it is essential to add overhead costs. These are important because they can account for about 200% or more of direct costs.

TABLE 4.6
Suggested design changes for controller

	Recommended design change	Items (from table 5)	Savings in time (seconds)
1	Plastic cover should be combined with frame and three screws should be eliminated and reoriented	19,20,21	54.35
2	Eliminate strap and two screws	6,7	27.9
3	Eliminate screws that hold PCB assembly	13	19.6
4	Eliminate need to reorient twice	4,16	18.00
5	Tube assembly and screw fastening operation should be eliminated	10,11	12.00
6	Earth lead should be eliminated	15	10.60
7	Connector should be eliminated	14	6.95

Source: Adapted from *Assembly Automation and Product Design* (Boothroyd).

The revised worksheet corresponding to suggested design changes is presented in Table 4.7.

The design principle of overconstraint leading to part redundancy is depicted pictorially in Figure 4.18.

4.5.5 A DfA Methodology Based on MTM Standards

An improved assembly methodology takes into consideration numerous factors such as weight, size and shape of components being disassembled, frequency of assembly tasks (based on the number of similar products being disassembled within a particular time frame), requirement of manpower, postural requirements, material handling requirements and need for component preparation. Several human factors in addition to design and economic factors merit consideration due to labor intensiveness of the assembly process.

The most common and widely used assembly operations are recorded and described in enough detail. Every disassembly operation is then subdivided into basic elemental tasks. It has been observed that only a fraction of all the tasks in the entire assembly operation are responsible for performing assembly. The remaining tasks constitute such actions as reaching for tools and grasping tools. An MTM-based second order index for assembly is presented in Table 4.8. The simplest assembly task of inserting an easily grasped object without the exertion of much force by hand by a trained worker under average conditions has been considered as the basic disassembly task. A score of 73 TMUs was assigned to this task which corresponded to the time duration of approximately 2 seconds. Subsequent scores were assigned based on the detailed study of most commonly encountered assembly operations.

Figure 4.19 depicts a system of measures that can be utilized to enhance ease of assembly of product architectures. Figure 4.20 depicts a methodology to enable

TABLE 4.7
Worksheet for modified controller assembly

Item	Number of Items	Manual handling code	Handling time per item in seconds	Manual insertion code	Insertion time per item in seconds	Total operation time in seconds	Figures for min parts	Description
1. Pressure regulator	1	30	1.95	00	1.50	3.45	1	Place pressure regulator in fixture
2. Plastic cover	1	30	1.95	06	5.50	7.45	1	Add plastic cover and hold down
3. Nut	1	00	1.13	39	8.00	9.13	0	Add nut and fasten screw
4. Knob assembly	1	30	1.95	08	6.50	8.45	1	Add knob assembly and fasten screw
5. Fastening screw	1			92	5.00	5.00		Fasten screw
6. Reorientation	1			98	9.00	9.00		Reorient assembly
7. Application of tape	1			99	12.00	12.00		Apply tape
8. Adapter nut	1	10	1.50	49	10.50	12.00	0	Add adapter nut and fasten screw
9. Sensor sub	1	30	1.95	39	8.00	9.90	1	Add sensor sub and fasten screw
10. PCB assembly	1	83	5.60	30	2.00	7.60	1	Add PCB assembly and snap fit
Total						**83.98**	**5**	

Source: *Assembly Automation and Product Design* (Boothroyd).

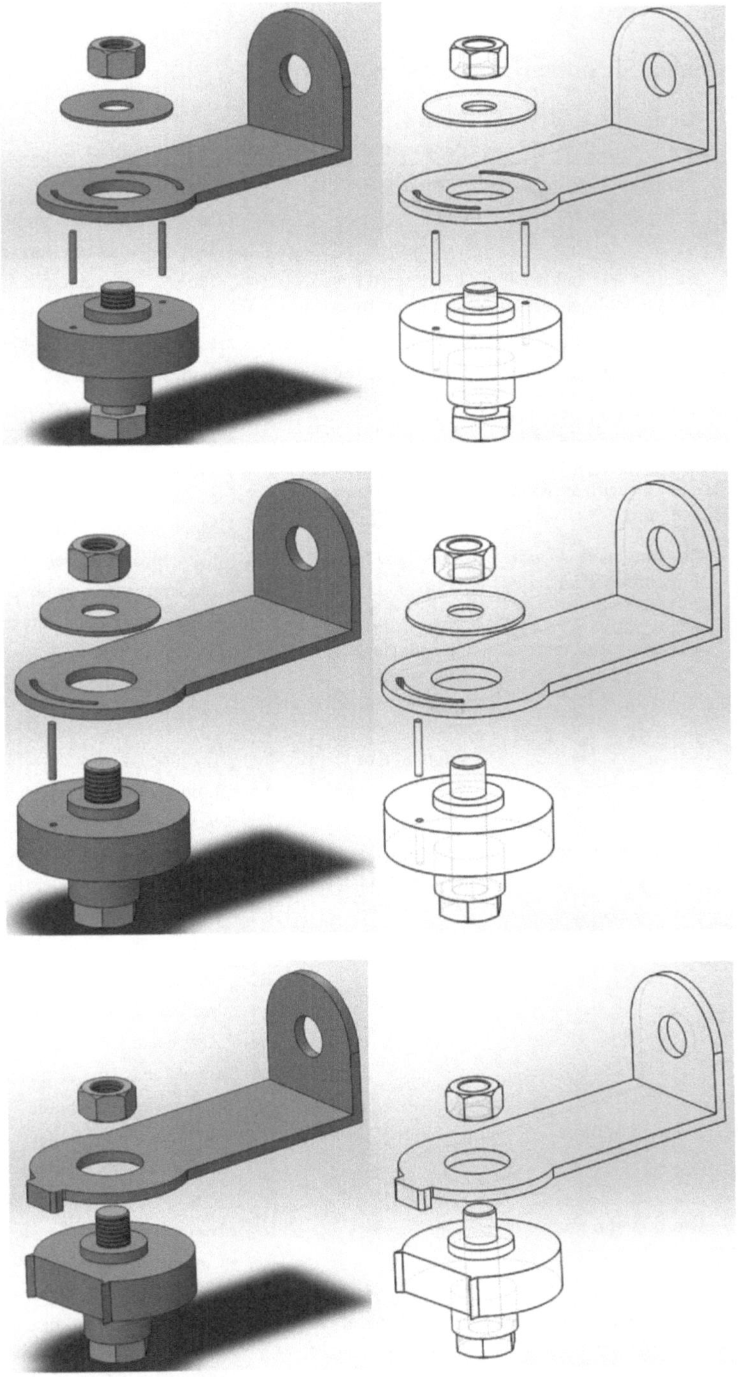

FIGURE 4.18 Design principle: Over constraint leading to part redundancy (Adapted from Boothroyd, 2005).

TABLE 4.8
Scoring system for numeric analysis of assembly

Design attribute	Design feature	Design parameters	Score	Interpretation
Assembly force	Straight line motion without exertion of pressure	Push operations by hand	0.5	Little effort required
			1	Moderate effort required
			2	Large amount of effort required
	Straight line and twisting motion without pressure	Twisting and push operations by hand	1	Little effort required
			2	Moderate effort required
			4	Large amount of effort required
	Straight line motion with exertion of pressure	Intersurface friction and/or wedging	2	Little effort required
			2.5	Moderate effort required
			4	Large amount of effort required
	Straight line and twisting motions with exertion of pressure	Intersurface friction and /or wedging	2.5	Little effort required
			3	Moderate effort required
			5	Large amount of effort required
	Twisting motions with pressure exertion	Material stiffness	2.5	Little effort required
			4	Moderate effort required
			6	Large amount of effort required
Material handling	Component/fastener size	Component dimensions (very large or very small)	2	Easily grasped
			3.5	Moderately difficult to grasp
			4	Difficult to grasp
		Magnitude of weight	2	Light (<7.5 lb)
			2.5	Moderately heavy (<17.5 lb)
			3	Very heavy (<27.5 lb)
	Component/fastener symmetry	Symmetric components are easy to handle	0.8	Light and symmetric
			1.2	Light and semisymmetric
			1.4	Light and asymmetric
			2	Moderately heavy, symmetric
			2.2	Moderately heavy, semisymmetric
			2.4	Moderately heavy, asymmetric
			4.4	Heavy and symmetric
			4.6	Heavy and semisymmetric
			5	Heavy and asymmetric
Requirement of tools for assembly	Exertion of force		1	No tools required
			2	Common tools required
			3	Specialized tools required
	Exertion of torque		1	No tools required
			2	Common tools required
			3	Specialized tools required
Accessibility of joints/ grooves	Dimensions	Length, breadth, depth, radius, angle made with surface	1	Shallow and broad fastener recesses, large and readily visible slot/recess in case of snap fits

TABLE 4.8
(Cont.)

Design attribute	Design feature	Design parameters	Score	Interpretation
			1.6	Deep and narrow fastener recesses, obscure slot/recess in case of snap fits
			2	Very deep and very narrow fastener recesses, slot for prying open snap fits difficult to locate
	Location	On plane surface	1	Groove location allows easy access
		On angular surface	1.6	Groove location is difficult to access; some manipulation required
		In a slot	2	Groove location very difficult to access
		On vertical surface	1.5	Some manipulation required against gravity
		On horizontal surface	1	Groove location allows easy access
Positioning	Level of accuracy required to position the tool	Symmetry	1.2	No accuracy required
			2	Some accuracy required
			5	High accuracy required
		Asymmetry	1.5	No accuracy required
			2	Some accuracy required
			5	High accuracy required

design improvement for product assembly. It should be noted that these tables and figures will be repeated in a subsequent chapters on designing for maintenance.

Scores depicted in Table 4.8 are computed using the basic MTM system. This system is used to compute time estimates for different operations such as lifting, inserting, locating and so on. Time estimates are based on statistical studies of respective operations. Table 4.9 illustrates application of MTM values to basic motions such as reach or move, position, grasp, disengagement, application of pressure as well as several body, leg and eye motions. Values are expressed in terms of time measurement units (TMUs). Remember that each TMU is worth 0.036 seconds. Thus, it is easy to estimate the assembly time for any assembly operation if it can be sufficiently broken down into its constituent elements. Table 4.9 is taken from the Maynard Handbook, Stegmerten, and Schwab (1948).

In order to perform an assembly operation manually, it is essential for an assembly worker to engage in a variety of tasks such as reaching for components and/or fasteners, grasping them, moving them to the work area, turning and applying pressure and so on. Tables 4.10 and 4.11 depict the TMU values for the functions: "Reach,"

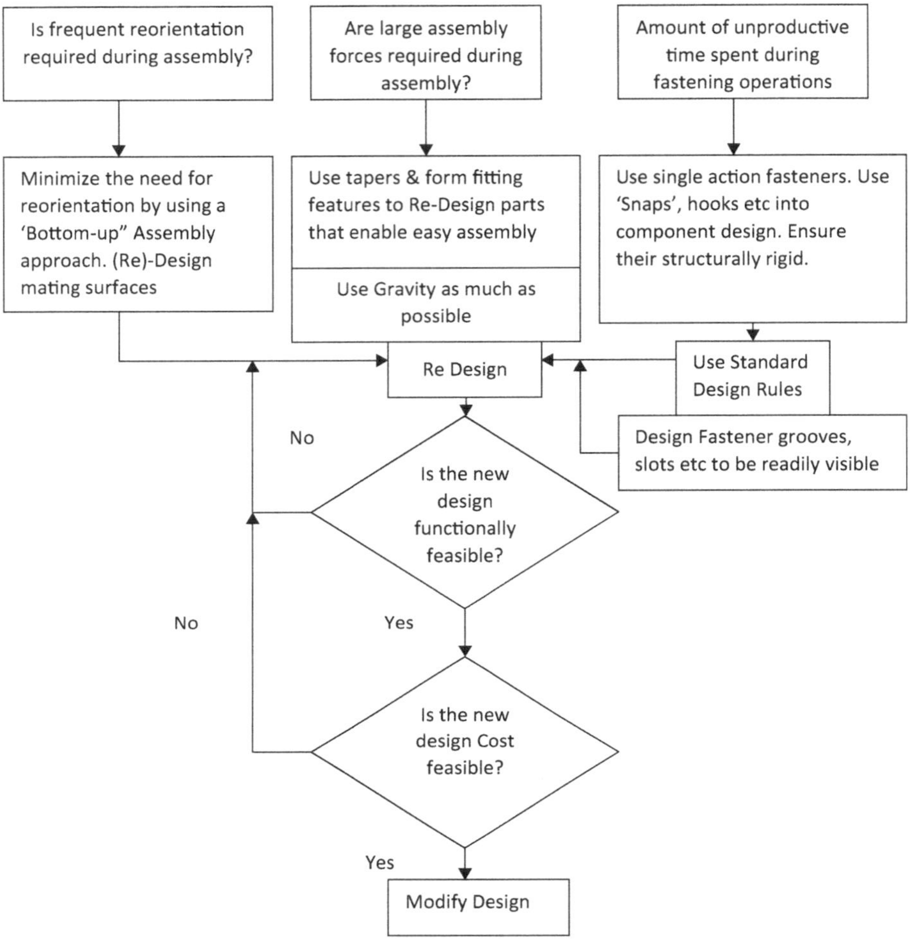

FIGURE 4.19 Enhancement of product design features to enable ease of assembly.

"move," "turn and apply pressure," "grasp" and "position" respectively. It will be noted that several design factors that dictate complexity of operations have already been accounted for in the formulation of the TMU values. Thus, a "regrasp" operation is likely to take longer to perform than an operation that involves picking up a medium or large object by itself that is capable of being easily grasped.

In order to move an object, it is necessary to work against gravity. Thus, design factors such as weight of the object as well as the amount of distance to be moved to affect assembly must be considered. This allowance has been included in computing TMU scores for the "move" function. TMU scores for the element: "Turn and apply pressure" are depicted in Table 4.12. It will be appreciated that most assembly functions that involve fasteners invariably include this particular element. Turning and applying pressure also involves a certain amount of skill and manual dexterity.

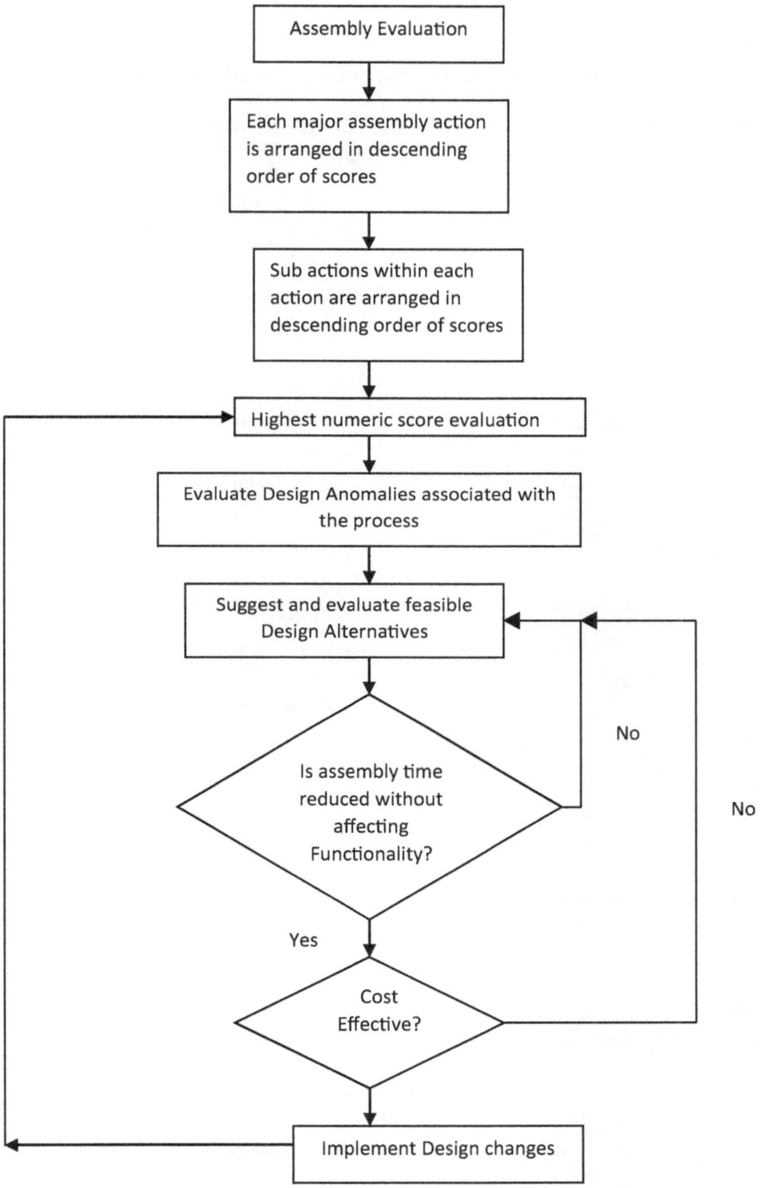

FIGURE 4.20 Methodology to enable design improvement for product assembly.

Note that the greater a screw needs to be turned (number of degrees is higher in this case), the longer it takes to perform the assembly operation.

An object that can be easily grasped is likely to entail less assembly time and vice versa. We have already dealt with this concept when we considered basic design

TABLE 4.9
Applying MTM values to basic manual motions involving assembly operations

Hand and arm motions			**Body, leg and eye motions**	
Reach or move		**TMU**		**TMU**
1"		2	Simple foot motion	10
2"		4	Foot motion with pressure	20
3"–12"		4+ length of motion	Leg motion	10
Over 12"		3+length of motion	Side step case 1	20
For "type 2 reaches" and "moves," use length of motions only			Side step case 2	40
Position			Eye time	10
Fit	**Symmetrical**	**Other**	Bend, stoop or kneel on one knee	35
Loose	10	15	Arise	35
Close	20	25	Kneel on both knees	80
Exact	50	55	Arise	90
Turn-apply pressure			Sit	40
Turn		6	Stand	50
Apply Pressure		20	Walk per pace	17
Grasp			1 TMU = 0.00001 Hour	
Simple		2	1 TMU = 0.0006 minutes	
Regrasp or transfer		6	1 TMU = 0.036 seconds	
Complex		10		
Disengage				
Loose		5		
Close		10		
Exact		30		

TABLE 4.10
Individual values for the "reach" function in the basic MTM system

Distance moved in inches	**Time (TMU)**				**Hand in motion**		
	A	**B**	**C or D**	**E**	**A**	**B**	**Description**
0.75 or less	2	2	2	2	1.6	1.6	A: Reach to object in fixed location or to object in another hand or on which other hand rests
1	2.5	2.5	3.6	2.4	2.3	2.3	
2	4	4	5.9	3.8	3.8	2.7	
3	5.3	5.3	7.3	5.3	4.5	3.6	
4	6.1	6.4	8.4	6.8	4.9	4.3	B: Reach to single object in location which may vary slightly from one cycle to another
5	6.5	7.8	9.4	7.4	5.3	5	
6	7	8.6	10.1	8	5.7	5.7	

TABLE 4.10
(Cont.)

Distance moved in inches	Time (TMU)				Hand in motion		
	A	B	C or D	E	A	B	Description
7	7.4	9.3	10.8	8.7	6.1	6.5	C: Reach to object jumbled with other objects in a group so that search and select occur
8	7.9	10.1	11.5	9.3	6.5	7.2	
9	8.3	10.8	12.2	9.9	6.9	7.9	
10	8.7	11.5	12.9	10.5	7.3	8.6	D: Reach to a very small object or where accurate grasp is required
12	9.6	12.9	14.2	11.8	8.1	10.1	
14	10.5	14.4	15.6	13	8.9	11.5	
16	11.4	15.8	17	14.2	9.7	12.9	E: Reach to indefinite location to get hand in position for body balance or next motion or out of way
18	12.3	17.2	18.4	15.5	10.5	14.4	
20	13.1	18.6	19.8	16.7	11.3	15.8	
22	14	20.1	21.2	18	12.1	17.3	
24	14.9	21.5	22.5	19.2	12.9	18.8	
26	15.8	22.9	23.9	20.4	13.7	20.2	
28	16.7	24.4	25.3	21.7	14.5	21.7	
30	17.5	25.8	26.7	22.9	15.3	23.2	

TABLE 4.11
Individual values for the "move" function in the basic MTM system

Distance moved in inches	Time (TMU)				Weight allowance			
	A	B	C	Hand in motion (B)	Weight (lbs) upto	Factor	Constant TMU	Description
0.75 or less	2	2	2	1.7	2.5	1.0	0	A: Move object to other hand or against stop
1	2.5	2.9	3.4	2.3				
2	3.6	4.6	5.2	2.9	7.5	1.06	2.2	B: Move object to approximate or indefinite location
3	4.9	5.7	6.7	3.6				
4	6.1	6.9	8	4.3	12.5	1.11	3.9	C: Move object to exact location
5	7.3	8	9.2	5				
6	8.1	8.9	10.3	5.7	17.5	1.17	5.6	
7	8.9	9.7	11.1	6.5				
8	9.7	10.6	11.8	7.2	22.5	1.22	7.4	
9	10.5	11.5	12.7	7.9				
10	11.3	12.2	13.5	8.6				
12	12.9	13.4	15.2	10	27.5	1.28	9.1	
14	14.4	14.6	16.9	11.4				

(continued)

TABLE 4.11
(Cont.)

	Time (TMU)				Weight allowance			
Distance moved in inches	A	B	C	Hand in motion (B)	Weight (lbs) upto	Factor	Constant TMU	Description
16	16	15.8	18.7	12.8	32.5	1.33	10.8	
18	17.6	17	20.4	14.2				
20	19.2	18.2	22.1	15.6	37.5	1.39	12.5	
22	20.8	19.4	23.8	17				
24	22.4	20.6	25.5	18.4	42.5	1.44	14.3	
26	24	21.8	27.3	19.8				
28	25.5	23.1	29	21.2	47.5	1.5	16	
30	27.1	24.3	30.7	22.7				

TABLE 4.12
Individual values for the "turn and apply pressure" function in the basic MTM system

	Time (TMU) for degrees turned										
Weight	30	45	60	75	90	105	120	135	150	165	180
0–2 pounds	2.8	3.5	4.1	4.8	5.4	6.1	6.8	7.4	8.1	8.7	9.4
2.1–10 pounds	4.4	5.5	6.5	7.5	8.5	9.6	10.6	11.6	12.7	13.7	14.8
10.1–35 pounds	8.4	10.5	12.3	14.4	16.2	18.3	20.4	22.2	24.3	26.1	28.2
Apply pressure case 1: 16.2 TMU											
Apply pressure case 2: 10.6 TMU											

principles for DfA. Individual TMU values for the "grasp" function are presented in Table 4.13.

It is essential to accurately position the component and/or fastener prior to applying force and turning. Thus, this is a time-intensive operation and must be accounted for. The individual TMU values for the "position" function are depicted in Table 4.14. Note that the ease with which a component/fastener can be positioned is a function of its symmetry (another design principle that we have already dealt with).

The "release" function accounts for the time taken to release appendages after the assembly function has been accomplished. Individual scores for this function are presented in Table 4.15. Given the lack of complexity, this is fairly straightforward.

The individual values for the "disengage" function are depicted in Table 4.16.

Individual values for the function "body, leg and foot motions" are presented in Table 4.17. This is an important aspect of some assembly processes and thus cannot be ignored.

TABLE 4.13
Individual TMU values for the "grasp" function in the basic MTM system

Case	Time (TMU)	Description
1A	2.0	Pick up grasp—Small, medium or large object by itself, easily grasped
1B	3.5	Very small object or object lying close against a flat surface
1C1	7.3	Interference with grasp on bottom and one side of nearly cylindrical object; diameter larger than 0.5 inches
1C2	8.7	Interference with grasp on bottom and one side of nearly cylindrical object; diameter 0.25–0.5 inches
1C3	10.8	Interference with grasp on bottom and one side of nearly cylindrical object; diameter less than 0.25 inches
2	5.6	Regrasp
3	5.6	Transfer grasp
4A	7.3	Object jumbled with other objects; search and select occurs; larger than 1" X 1" X 1"
4B	9.1	Object jumbled with other objects; search and select occurs; larger than 1" X 1" X 1"
4C	12.9	Object jumbled with other objects; search and select occurs; smaller than 0.24" X 0.25" X 0.125"
5	0	Contact, sliding or hook grasp

TABLE 4.14
Individual values for the "position" function in the basic MTM system

Class of fit		Symmetry	Easy to handle	Difficult to handle
1: Loose	No pressure required	S	5.6	11.2
		SS	9.1	14.7
		NS	10.4	16.0
2: Close	Light pressure required	S	16.2	21.8
		SS	19.7	25.3
		NS	21.0	26.6
3: Exact	Heavy pressure required	S	43.0	48.6
		SS	46.5	52.1
		NS	47.8	53.4

TABLE 4.15
Individual TMU values for the "release" function in the basic MTM system

Case	Time (TMU)	Description
1	2.0	Normal release performed by opening fingers as an independent motion
2	0	Contact release

TABLE 4.16
Individual values for the "disengage" function in the basic MTM system

Class of fit	Easy to handle	Difficult to handle
1: Loose—Very slight effort, blends with subsequent move	4.0	5.7
2: Close—Normal effort, slight recoil	7.5	11.8
3: Tight—Considerable effort, hand recoils markedly	22.9	34.7

TABLE 4.17
Individual values for the "body, leg and foot motions" function in the basic MTM system

Description	Symbol	Distance	Time (TMU)
Foot motion—hinged at ankle	FM	Upto 4"	8.5
with heavy pressure	FMP	Upto 6"	19.1
leg or foreleg motion	LM	Each additional inch	7.1
			1.2
Sidestep—	SS-C1	Less than 12"	Use REACH or
Case 1: Complete when leading	SS-C2	12"	MOVE time
leg contacts floor		Each additional inch	17.0
Case 2: Lagging leg must contact		12"	0.6
floor before next motion can be		Each additional inch	34.1
made			1.1
Bend, stoop or kneel on one knee	B,S,K O K		29.0
Arise	AB,AS,AK,OK		31.9
Kneel on floor:	KBK		69.4
Both knees	AKBK		76.7
Arise			
Sit	SIT		34.7
Stand from sitting position	STD		43.4
Turn body 45–90°	TBC 1		18.6
Case 1: Complete when leading	TBC 2		37.2
leg contacts floor			
Case 2: lagging leg must contact			
floor before next motion can be			
made			
Walk	W-FT	Per foot	5.3
Walk	W-P	Per pace	15.0

It is clear that once design anomalies are identified, design alterations can be undertaken in order to improve product design for the assembly perspective. Once corresponding design decisions have been made, the cost of manufacturing those components can be taken into consideration in order to optimize the manufacturing process and maximize profit potential.

A case study to illustrate the practical application of the DfA methodology has been presented. The goal is to redesign the bottom panel of a laptop computer. The

TABLE 4.18
Components of the bottom panel of a laptop computer

Component name	Component material	Quantity
Bottom panel	Aluminium	1
Access hatch	ABS Plastic	1
Fastening screws	Aluminium	18
Main body of the laptop	Various	1

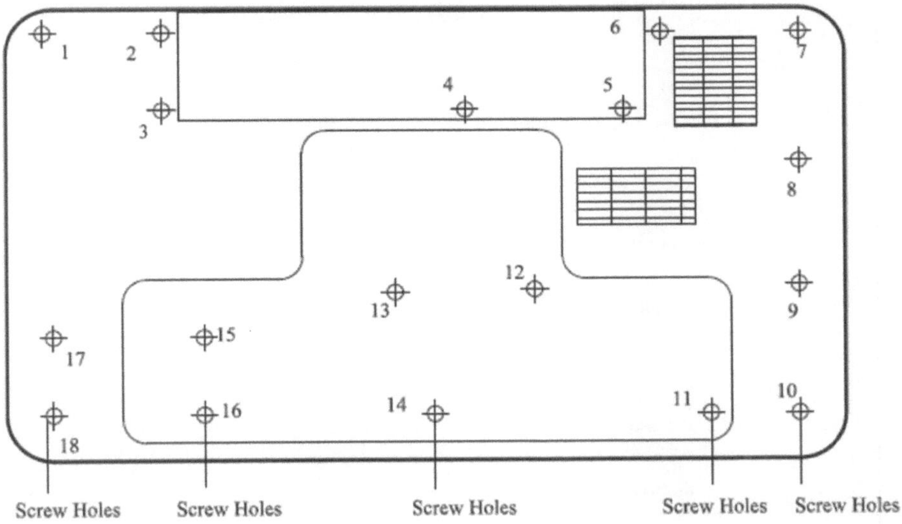

FIGURE 4.21 Bottom panel of a laptop computer to be evaluated for ease of assembly.

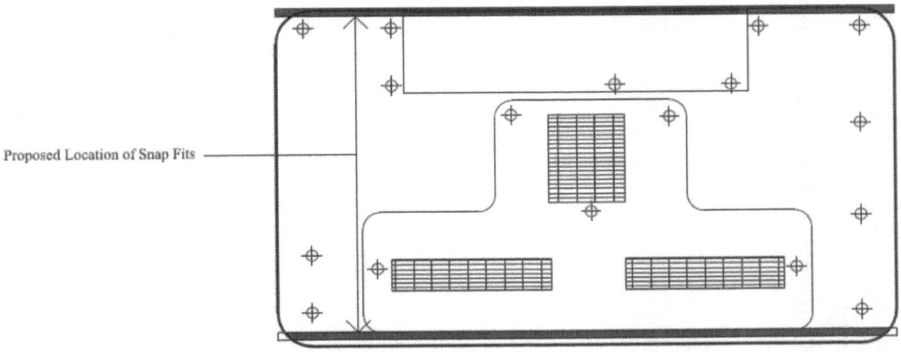

FIGURE 4.22 Design modification of bottom panel using DfA analysis.

TABLE 4.19
Assembly operation of a computer monitor

Task no.	Task description * Assembly of computer Monitor*	Task total	Assembly force			Material handling			Tooling		Accessibility and positioning			Allowances			
			Intersurface friction	Intersurface wedging	Material stiffness	Component size	Component weight	Component symmetry	Force exertion	Torque exertion	Dimensions	Location	Accuracy of tool placement	Posture allowance	Motions allowance	Manpower allowance	Visual fatigue allowance
1	Assemble bottom panel																
1a	Place bottom panel in place	10.92	–	2	–	2	2	1.2	1	–	1	–	1.2	–	–	–	5%
1b	Screw first of 18 screws	15.12	2	–	–	2	2	0.8	–	2	1.6	2	2	–	–	–	5%
1c	Screw second of 18 screws	15.12	2	–	–	2	2	0.8	–	2	1.6	2	2	–	–	–	5%
1d	Screw 3rd of 18 screws	15.12	2	–	–	2	2	0.8	–	2	1.6	2	2	–	–	–	5%
1e	Screw 4th of 18 screws	15.12	2	–	–	2	2	0.8	–	2	1.6	2	2	–	–	–	5%
1f	Screw 5th of 18 screws	15.12	2	–	–	2	2	0.8	–	2	1.6	2	2	–	–	–	5%
1g	Screw 6th of 18 screws	15.12	2	–	–	2	2	0.8	–	2	1.6	2	2	–	–	–	5%
1h	Screw 7th of 18 screws	15.12	2	–	–	2	2	0.8	–	2	1.6	2	2	–	–	–	5%

1i	Screw 8th of 18 screws	15.12	2	–	–	2	2	0.8	–	2	1.6	2	2	–	–	–	5%
1j	Screw 9th of 18 screws	15.12	2	–	–	2	2	0.8	–	2	1.6	2	2	–	–	–	5%
1k	Screw 10th of 18 screws	15.12	2	–	–	2	2	0.8	–	2	1.6	2	2	–	–	–	5%
1l	Screw 11th of 18 screws	15.12	2	–	–	2	2	0.8	–	2	1.6	2	2	–	–	–	5%
1m	Screw 12th of 18 screws	15.12	2	–	–	2	2	0.8	–	2	1.6	2	2	–	–	–	5%
1n	Screw 13th of 18 screws	15.12	2	–	–	2	2	0.8	–	2	1.6	2	2	–	–	–	5%
1o	Screw 14th of 18 screws	15.12	2	–	–	2	2	0.8	–	2	1.6	2	2	–	–	–	5%
1p	Screw 15th of 18 screws	15.12	2	–	–	2	2	0.8	–	2	1.6	2	2	–	–	–	5%
1q	Screw 16th of 18 screws	15.12	2	–	–	2	2	0.8	–	2	1.6	2	2	–	–	–	5%
1r	Screw 17th of 18 screws	15.12	2	–	–	2	2	0.8	–	2	1.6	2	2	–	–	–	5%
1s	Screw 18th of 18 screws	15.12	2	–	–	2	2	0.8	–	2	1.6	2	2	–	–	–	5%
		283.08				Total score for assembly operation. Total time = 1.70 minutes.											

process is demonstrated in Table 4.19. The various components of the bottom panel are depicted in Table 4.18.

The bottom panel in its current state is depicted in Figure 4.21.

The part redesign using snap fits is depicted in Figure 4.22.

It is clear from Table 4.19 that the total amount of time taken to assemble a typical bottom panel of a laptop computer is 1.70 minutes. Note that the incorporation of snap fits in highlighted locations is likely to reduce the assembly time significantly.

4.6 CONCLUSION

In this chapter we examined the different types of assembly processes that are based on production volume. Specifically, the processes of manual assembly, robotic assembly and automatic assembly were dealt with. In addition, design guidelines to facilitate part (re)design were also presented. Three design methodologies in particular were presented (Hitachi, Lucas and Boothroyd) in addition to another that used MTM scores to estimate assembly time. Product DfA is important irrespective of whether the end goal is to design a sustainable product. Even a product that is being designed to be eventually disposed of in a landfill after just one use (and not for reuse, recycling or remanufacturing) needs to be put together in a coherent manner such that it is able to perform its intended function satisfactorily. Designing a product for ease of assembly is extremely important so much so that one of the original DfX paradigms was known as DfMA.

REFERENCES

Boothroyd, G. (2005). *Assembly Automation and Product Design*. Second Edition. Taylor and Francis-CRC Press. Boca Raton, FL.

Huang, G.Q. (1996). *Design for X: Concurrent Engineering Imperatives*. Springer, Dordrecht, Chapman and Hall.

Maynard, H.B., Stegmerten, G.J. and Schwab, J.L. (1948). *Methods—Time Measurement*. McGraw Hill, Pittsburgh, PA.

5 Designing for Maintenance

5.1 WHAT IS MAINTENANCE?

Sustainable machines and equipment are required to last longer than their nonsustainable counterparts. It should also be easier to repair and remanufacture such machines. Thus, it is crucial to design a machine for ease of maintenance if the ultimate goal is to make it more sustainable.

Every system or product requires routine maintenance. Whether it is a mechanical system, an electrical system, an electromechanical system or a chemical system, it must be maintained periodically. Even the human body and mind is no exception to this rule. A system or product that is not maintained regularly eventually falls into disrepair. Think what would happen if you do not change the engine oil in your car regularly. The consequences would be truly disastrous if worn brake rotors were not changed when required. Lack of maintenance leads to loss of function whereas regular maintenance can lead to extending the product's function. Maintenance can be achieved through a host of actions such as repair, overhaul, replacement or by automated action.

Significant amount of investment both in terms of human resources and monetary investment is required in maintaining modern systems/equipment. This is mainly due to their increasing complexity. Generally, the more complex the system, the more difficult it is to maintain. Thus, designers should strive to improve the maintenance friendliness of new equipment. This can be facilitated by incorporating maintenance analysis during the design, acquisition and selection phase of the new product/equipment. If machines/products are not designed for ease of maintenance (i.e., taking future maintenance requirements into account), higher costs and times need to be incurred even for a simple replacement operation (Hubka, 1994). Thus, maintenance needs to be considered as a design attribute early during the product design phase. As a matter of fact, this attribute is so important that its early consideration leads to improvement in implied quality of the product (Di Gironimo et al., 2004).

Reliability is defined as the ability of a machine/system to work successfully over a prolonged period of time. Very little practically applicable results have been obtained by designers despite their ongoing obsession and fascination with improving system reliability. A system that is hundred percent reliable is nothing more than a mirage. The benefits of periodically maintaining equipment are many. Doing so by adherence to a strict maintenance regimen helps prolong the life of an equipment. Future breakdowns are also avoided, and smooth operation of the equipment is ensured.

Morgan et al. (1963) have defined maintainability as

> the degree of facility with which an equipment or system is capable of being retained in, or restored to, serviceable operation. It is a function of parts accessibility, interval configuration, use and repair environment and the time, tools and training required to effect maintenance.

Furthermore, according to the department of defense, maintainability is defined as "a characteristic of design and installation which is expressed as the probability that an item will conform to specified conditions within a given period of time when maintenance action is performed in accordance with prescribed procedures and resources" (Harring and Greenman, 1965). It should be clear by now that designing for maintenance plays a critical role in ensuring reliable operation of equipment. It can be concluded with a significant degree of confidence that effective maintenance leads to reliability and not vice versa.

Both commercial equipment and military equipment need to be maintained. As far as a commercial product is concerned, success in the marketplace depends on whether such a product can be maintained or returned to usable condition within a reasonable period of time. It is also important that the maintenance operation is not too expensive.

As far as military systems are concerned, it should be obvious that the competition is not between companies but between countries. Deterrence of aggression enables a country to survive and thrive. If deterrence is not possible, then subjugation of the enemy is the only alternative. In light of these facts, the defense industry considers maintainability as one of the more important parameters that directly contributes to materiel readiness (Harring and Greenman, 1965).

To sum up our discussion, fatigue failure is a fact of life. All physical components will eventually fail. The machine assembly and its components will suffer breakdown as a result of fatigue and wear. This process takes time. It can also occur as a result of improper use (please refer to the basic concept of design review delineated in chapter 1 to gain a better understanding of this concept). Also, it is not possible to circumvent the need for periodic maintenance simply by incorporating redundancy into equipment design.

5.2 WHAT ARE THE FACTORS THAT AFFECT MAINTAINABILITY?

Over the past few decades the rate of technological evolution has skyrocketed. It has resulted in equipment design that continues to become increasingly complicated. Consider a very simple example. Think about the presence of sensors and complicated electronics in every machine. There used to be a time when a car engine used to consist solely of mechanical components. As such, it could be repaired easily. Those days are long gone. Today's car engines are primarily controlled by electronic circuits and sensors. This has made the repair process infinitely more complicated and as a result, more expensive. The growing degree of machine complexity has far outpaced evolution in reliability technologies (Crawford and Altman, 1972; Morgan et al., 1963; Oborne, 1981). There has been a proliferation in new problems in equipment downtime. This implies that machines are beginning to fail more frequently, and the repair process is beginning to become prohibitively expensive. Periodic maintenance is the only feasible tool that can serve to reduce machine downtime. For this reason, it has assumed growing importance in modern times (Imrhan, 1991).

Historically, designing equipment for ease of maintenance has been practiced as an art. Designers have relied upon a host of "hunches" to accomplish their objective. It has not been the result of scientific investigation, rather the "hunches" that have resulted from common sense and field experience (Oborne, 1981). Maintenance could probably be characterized as one of the more expensive of all human-machine systems. This is attributable to the following factors:

- For the reasons that we have seen before, the need to perform maintenance activities continues to increase almost exponentially.
- Maintenance is a manual activity. It needs to be performed by a well-trained labor force. The cost of manual labor has continued to increase over the years. Given the fact that maintenance is perhaps one of the only fields of industrial operation that relies solely on human capital and skill, it is important to accurately estimate the cost of human labor.

The military aircraft industry is rife with examples that can help corroborate the above claim. Some of these examples are presented as follows:

- According to McDaniel and Askren, aircraft maintenance costs in the US were estimated to amount to approximately 35% of life cycle costs of military systems in 1985.
- Technically, modern aircraft have become increasingly more complex. As a result, they can no longer be maintained quickly and cost effectively. One of the reasons for this is that many parts are designed using CAD systems. These systems are capable of designing intricate shapes and patterns that are not easy to maintain (Adams, 1988).
- As far back as 1970, as much as a quarter of the budget of the US department of defense was allocated to pay for equipment maintenance costs (Smith et al., 1970). The increasing level of complexity of components and machine assemblies as outlined here has only prompted this fraction to continue to grow.

However, this problem can be solved. The solution lies in designing machine and equipment so that they are easy to maintain. Such machines lend themselves more easily to being maintained quickly and less expensively. Maintenance operations can now be performed at a fraction of their regular cost. They also take less time. The importance of designing equipment for ease of maintenance can thus not be overstated. In order to better understand the maintenance function itself, functional design (also referred to as design for operability) needs to be clearly distinguished from design for maintenance. This is presented as follows:

It is important to design for maintenance. However, the task is easier said than done. This is attributable to a variety of reasons:

- From the environmental point of view, the workplace where the process of maintenance is carried out involves a great degree of variability and unpredictability. Limited and cramped workspaces that have not been designed to perform such operations are characteristic of the type of environment in which maintenance technicians are forced to work.
- There is a staggering degree of variability in equipment that needs to be maintained. Consumer electronics products are a prime example. Rapidly evolving technology leads to faster obsolescence of such products. The maintenance crew has no option but to train and retrain frequently in order to keep pace with technological evolution.
- Equipment operability and maintainability constitute mutually conflicting design goals. For instance, machine maintenance is enhanced by provision of clearance within a machine and between machine parts. This design practice does not always lend itself to improving the functionality of the machine.
- Product design is being driven increasingly by the trend of miniaturization (Tichauer, 1966). Miniaturization leads to reduced production cost. Machines that follow this practice are often easier to manipulate during operation and transportation. It also satisfies the demands of a workplace that has been gradually shrinking. A smaller machine can fit easily into a smaller workplace.

The rate of production and safety can be enhanced by maintenance-friendly machines. According to Ferguson et al. (1985), it is less probable that machines that are not easy to maintain routinely will receive the required standard of maintenance. For instance, machines tended to suffer breakdown more often as a direct result of lack of maintenance or abuse of equipment and not just because they were incorrectly engineered (Johnson, 1988). Significant strides have been made within the field of reliability. Despite this, it is highly unlikely that a totally reliable and cost feasible machine will be developed. Thus, it is always advisable to practice good maintainability.

Maintenance-friendly machines are the result of a design activity. Most maintenance-related research focuses on mathematical algorithms. These algorithms seek to schedule maintenance activities that can be performed with the least equipment downtime.

A design focus on maintenance-friendly machines is likely to solve many of the aforementioned problems. Given that machine maintenance is a largely manual process, a design methodology focusing on simplifying this process will incorporate ergonomic data. Ergonomics data that are available in texts (e.g., Van Cott and Kinkade, 1972) need to be adapted so that they can help designers achieve a tradeoff between ergonomics and engineering issues. This chapter will present some design approaches and guidelines that do exist on this topic.

5.3 ELEMENTS OF THE MAINTENANCE OPERATION

The maintenance concepts and requirements for any system are described by maintenance elements. The analysis and verification of customer requirements is included

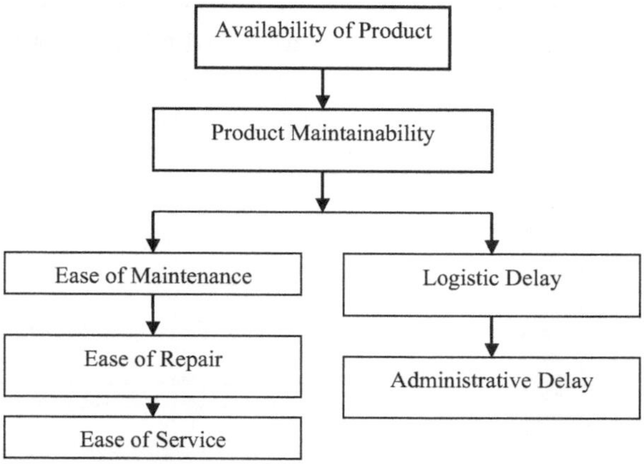

FIGURE 5.1 Relationship between customer satisfaction and the maintenance function (Adapted from Reiche, 1994).

in this. How each element is prioritized depends upon specific customer requirements. Individual elements and interconnections between them are depicted in Figure 5.1. These elements should be studied carefully in order to enhance system maintainability after it has been conceptualized. This is accomplished by fully integrating various elements that are part of the initial tasks that need to be performed. The system design should include each of these elements, which in turn should be carefully controlled (Reiche, 1994). Each element should be implemented on time and not lag the design of the system.

The idea of customer satisfaction as a measure of reliability and maintainability has been promoted by the International Electrotechnical Commission (IEC). Figure 5.1 depicts the maintenance parameter. Each element is subclassified in Figure 5.2.

Corrective and/or preventive maintenance procedures need to be carried out in order to maintain a system/equipment with minimum downtime. It is also preferable of such procedures to use minimal amount of resources. Skilled manpower, tools, test equipment, technical expertise and materials are examples of resources necessary to perform maintenance procedures. Some basic concepts related to designing a product for ease of maintenance are outlined in the following section. Many of these concepts are maintenance philosophies in themselves. They can be used as fundamental building blocks that would eventually lead to a cohesive design for maintenance methodology.

5.4 TYPES OF MAINTENANCE PROCEDURES

5.4.1 Corrective Maintenance

Corrective maintenance is performed after a machine or equipment has suffered breakdown. The failure of the product could be attributable to any number of reasons ranging from lack of regular maintenance to use to improper material to operator error. A restorative action is necessary to reinstate the operability of the equipment after

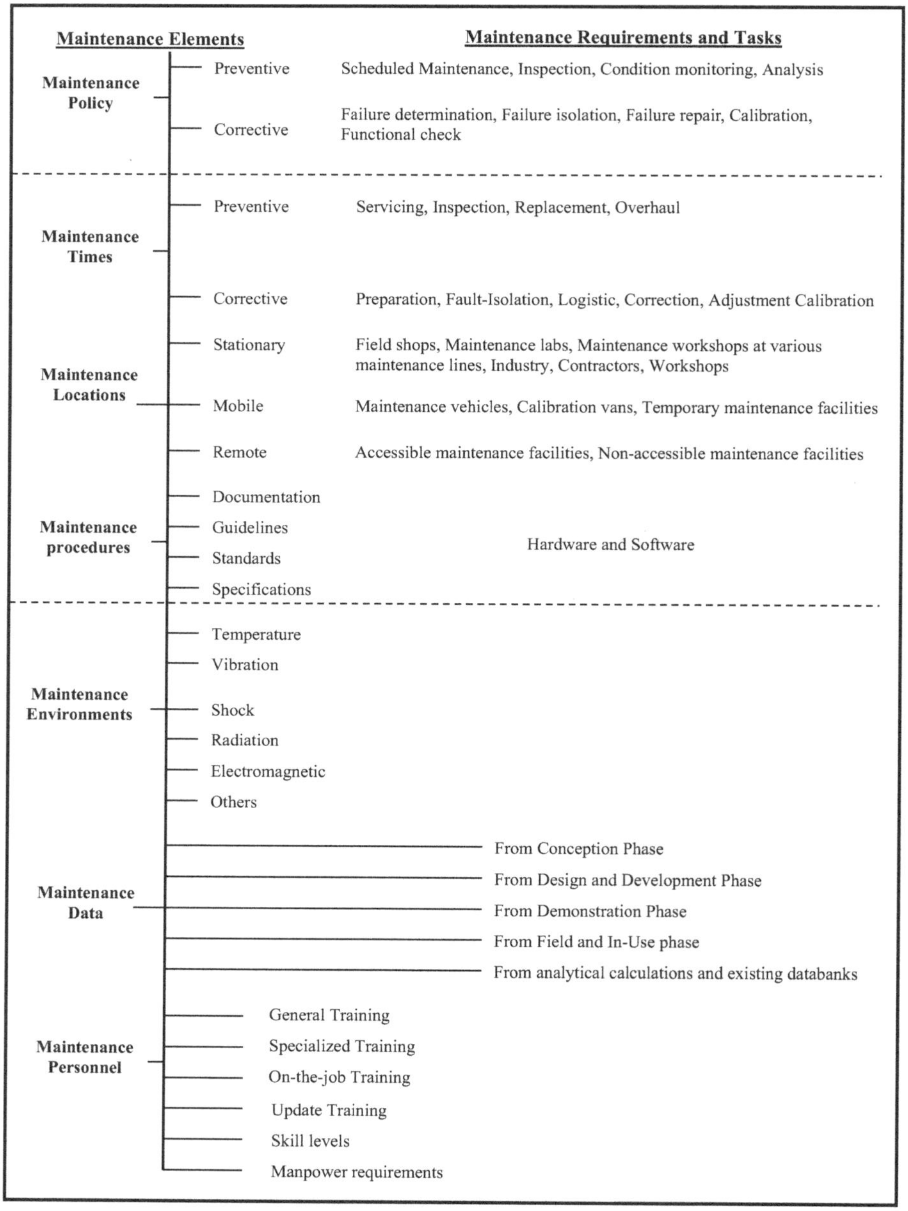

FIGURE 5.2 Different elements of the maintenance operation are interrelated (Adapted from Reiche, 1994).

each failure. Absent such action, it would be impossible for the machine to resume normal production. Corrective maintenance is performed using the following steps:

- Detecting a failure is not enough. It is necessary to objectively confirm it. The item is usually returned to service if confirmation is not possible. This type of situation wastes a significant amount of time and cost. Unnecessarily large amounts of inventory are required to be carried all the time because of it.
- Confirmation of the failure, on the contrary, implies that corrective action needs to be taken. This implies that the machine must be repaired. This is accomplished by preparing the item for maintenance. A failure report is also completed simultaneously.
- The next step consists of localizing and isolating the failed part in the assembly. Addressing this root cause of failure will help restore the machine to working condition.
- The failed part can be either disposed of or repaired. In the former case, a new part is installed in its place. Broken connections, open circuit board on a printed circuit board (PCB) or a poor solder are examples of parts/connections that can be repaired.
- After it has been repaired, the item is then reassembled, realigned and adjusted. It is checked thoroughly before it is put back to use.

There is a large amount of uncertainty associated with corrective maintenance. This is its primary disadvantage. It will also be appreciated that due to the reactive nature of this procedure, a single failure is capable of shutting down an entire production line. This in turn can cause problems such as bottlenecking with respect to material flow, not to mention lost productivity.

Many of the problems associated with corrective maintenance can be solved by using predictive maintenance. The basic gist of this type of maintenance is that it is proactive in nature. It does not wait for equipment to breakdown before maintenance procedures are carried out. Rather, equipment is routinely taken offline at regular intervals so it can be maintained. This practice prevents (or postpones) its breakdown.

5.4.2 Predictive Maintenance

The primary goal of predictive maintenance is to minimize the probability of equipment failure. It is also referred to as preventive maintenance. As its name suggests, this procedure is designed to prevent machine breakdown. It is also sometimes referred to as use-based maintenance (Swanson, 2001).

Predictive maintenance consists of maintenance activities that are performed at regular intervals of time. It can also be performed after a particular amount of equipment use (Gits, 1992; Herbaty, 1990). The specific time interval at which maintenance activities are undertaken is determined by the estimated probability of equipment failure during that interval. Equipment lubrication, parts replacement, cleaning and adjustment (e.g., tightening, slackening, etc.) are examples of preventive maintenance tasks. Maintenance personnel could also check equipment for telltale signs of deterioration as an integral part of the preventive maintenance process.

It will be apparent to the reader that since preventive maintenance is carried out at regular intervals of time, it follows a predetermined schedule. This is necessary if the practice is to be fully effective. As a matter of fact, many systems/equipment have a preventive maintenance schedules routinely published for them. This practice is fine for older equipment. However, as far as new equipment is concerned, the manufacturer provides information to the customer and maintenance schedules are established based on such information. Decisions concerning maintenance schedules and procedures are greatly influenced by preventive maintenance (Reiche, 1994). It is essential to periodically inspect equipment if preventive maintenance is to be truly effective.

The effectiveness of preventive maintenance is enhanced by preparing and adhering to a so called preventive maintenance plan. The objectives of such a plan should be clear. The following are examples of such objectives:

- Maintain the design reliability and availability of a system.
- Perform preventive maintenance so that the need for corrective maintenance actions is reduced.
- Improve the effectiveness of maintenance and increase the amount of planned maintenance.

Standard practices such as predicting maintenance actions, applying diagnostic procedures to detect system deterioration prior to failure, regular calibrations and inspections, regularly monitoring system performance as well as making repairs and overhauls based on test results can help accomplish these goals (Reiche, 1994).

Despite its myriad advantages, equipment still needs to be taken offline in order to perform preventive maintenance. Thus, it is necessary to plan for machine downtime. It can therefore be construed that the many advantages that can accrue as a result of preventive maintenance can be obtained only at the cost of machine downtime.

Predictive maintenance is an adaptation of the standard preventive maintenance strategy outlined earlier. It is based on similar principles. The exception is that the need for specific maintenance actions is determined by a different set of criteria. This method uses different types of diagnostic tools to measure the physical condition of equipment. Examples of such conditions include physical attributes such as temperature, vibration, noise, corrosion and lubrication (Eade, 1997). These can be likened to symptoms of a disease. They are not related to inherent material properties. The system is taken offline to rectify the problem if and when any combination of these indicators registers a specified level.

Predictive maintenance does not need equipment to be regularly taken offline. It is done only when the need to do so is imminent. How do we know when it is imminent? Like we have mentioned before, equipment will "tell you" when it needs to be maintained by registering specific "symptoms" of its condition. It is a case of a simple IF/THEN condition. This constitutes a very big advantage (Herbaty, 1990; Nakajima, 1989).

In summary, preventive maintenance is routinely performed so that the following goals may be achieved (Smith and Hinchcliffe, 2004):

- Prevent equipment from failing. If it does fail, mitigate the effects of that failure.
- Detect the probability of imminent failure. The maintenance engineer can take precautionary actions to prevent a potential catastrophic failure.
- Not all failures are readily obvious. Failures that are hidden can also be detected.

5.4.3 Maintaining a Degrading System

Systems degradation is a fact of life. Every physical system wears down over time. If this wear is allowed to continue unabated, it will eventually get out of hand and cause catastrophic failure. Such an eventuality can be avoided by resorting to preventive maintenance. Periodic review of the maintenance actions that need to be taken can achieve this objective. The maintenance schedule can be optimized by monitoring the degradation of the system ahead of time. Factors such as maintenance costs and distribution of degradation can be added by doing so (Reiche, 1994). The amount of degradation or wear is measured after each monitoring period. Maintenance is carried out if too much wear is noticed. It is obvious that the amount of degradation is a positive entity (because there is no such thing as "negative wear"). It is also a continuous random variable. It is about the same for each monitoring period and is independently distributed. A maintenance plan yields optimal results when it includes cost factors. Thus, it is necessary to include the costs of overhauling and operating a system in the evaluation.

5.4.4 Aggressive Maintenance

The goal of an aggressive maintenance strategy is to improve overall equipment operation. As such, it takes the concept of preventive maintenance a step further. The concept of total productive maintenance (TPM) has inspired the idea of aggressive maintenance. Let us gain a better understanding of TPM in order to fully understand aggressive maintenance.

TPM is a philosophy that chiefly deals with maintenance management designed in order to complement the implementation of just-in-time (JIT) systems in Japanese plants (Swanson, 2001). It utilizes a partnership approach to the practice of maintenance (Maggard and Rhyne, 1992). The objective of TPM activities is to eliminate the "six major losses" related to equipment maintenance.

These losses are enumerated as follows (Macaulay, 1988):

- Equipment failure.
- Set-up time and adjustment time.
- Equipment idling and minor stoppages.
- Reduced speed.

- Defects in process.
- Reduced manufacturing yield.

The concept of TPM utilizes small groups or teams that create a cooperative relationship between maintenance and production. Maintenance tasks are facilitated by this relationship. The team approach also requires production workers to get involved in performing maintenance tasks. This allows them a role in monitoring equipment and maintaining it. The practice of TPM results in refining the skill of production workers and improving their efficiency in maintaining equipment.

The main objective of maintenance prevention teams is to improve the design of equipment so that its performance can in turn be improved (Swanson, 2001). In order to achieve this goal, in keeping with the principles of concurrent product design, the engineering department and maintenance department work together during the early stages of the product design process. Equipment that is designed in such a way is easy to operate and maintain (Adair-Heeley, 1989; Goto, 1989).

Improved equipment availability and reduced maintenance costs are the main advantages of TPM and aggressive maintenance. This results in reduced repair time and better maintenance efficiency.

5.5 PLANNING FOR MAINTENANCE AND ITS MANAGEMENT THROUGH DESIGN REVIEW

A comprehensive design methodology should take into consideration various facets of design such as functionality, usability, assembly disassembly, maintenance and so on. Just because this chapter deals with the topic of maintenance it does not imply that it should be the sole aim of any design activity. Designing a product for maintainability is just one more goal of the design process. The design will by no means be ineffective or fail to perform its intended function if it is not designed for ease of maintenance (this is obviously the case in many equipment designs on the market today). However, a machine/product that is designed so that it is easy to maintain is likely to last longer and thus be more sustainable.

Good maintainability and reliability is imparted to a product by means of a design review. The process consists of "a quantitative and qualitative examination of a proposed design to ensure that it is safe and has optimal performance with respect to maintainability, reliability and performance variables needed to specify the equipment" (Thompson, 1999).

Design reviews are undertaken at four principal levels of design. They are listed as follows:

- Review of design specifications. This includes reviewing market need in product design.
- Systems review.
- Evaluation of equipment and functional units.
- Component analysis.

Subsystems should also be included in the system-level review. Also, equipment review should include subassemblies. A commonsense approach is used to classify different parts of a machine/product. In most cases, it is enough to include the aforementioned four levels of classification. This enables the following:

- Selecting appropriate review methods for different tasks.
- Adopting a systematic approach to enable an efficient and effective design review.

A brief description of each activity in the design review is presented in Table 5.1.

5.5.1 Reviewing Design Specifications at the Initial Stage

The design specification review process ensures that details regarding all parts and specifications are thoroughly comprehended at the very beginning of the process. It also ensures that the importance of different statements is appreciated. If any miscommunication regarding design specification exists between the design team and the client, it can be addressed during this stage of the design process. The most important and defining characteristics of the design are discussed in order to accomplish this objective. Most contractual disputes, if they were to occur, tend to revolve

TABLE 5.1
Structured design review procedure

Stage	Activity	Purpose	Timing
1.	Review of design specifications	To ensure that the significance of all points contained within the design specification is understood	Prior to the commencement of any design activity
2.	Activity systems level review	To identify the critical areas of the design that may affect plant availability, and to communicate the details to the design teams the necessity to pay particular attention to these areas	Prior to the start of equipment design
		To comment on the advisability of pursuing projects with a high-risk content	
		To examine equipment groups to maximize uniformity and stability	After the completion of the first equipment designs
		To maximize the reliability systems formed by manufacturing and process considerations	
3.	Equipment (functional unit) evaluation	To evaluate quantitatively critical items of equipment	After the completion of the first detailed designs
		To undertake qualitative reviews of equipment	
4.	Component analysis	To check that certain important sets of components will not give rise to maintainability or reliability problems in service	After the completion of the first detailed design

Source: Adapted from Thompson (1999).

around design specifications. Thus, it is crucial for all parties concerned to thoroughly understand this basic aspect of the design. As far as maintainability is concerned, the following design specifications are most important:

- It is important to define and understand that the objectives of maintainability and reliability are quantitative in nature. This realization helps avoid any miscommunication in the future. Take for example a design specification that says that an access hatch needs to be located 5 ft from the ground. This is an example of an objective design specification related to maintainability and no one can possibly misunderstand it.
- The twin factors of maintainability and reliability are often influenced by environmental conditions. It is essential to thoroughly consider and address such conditions.
- It is important that any attributes related to maintainability are given careful and thorough consideration. Pertinent examples of such attributes include breaking down the design into modules (modular construction), carefully assessing the skill level of the maintenance crew and designs that would require the crew to possess multiple sets of skills.
- The equipment design should objectively demonstrate that the resulting machine can be properly maintained. It should also be easy to carry out the maintenance activity. Thus, reliability is an important consideration. Specifications should also include information on what are the acceptance criteria for the process.

5.5.2 System Review before Delving into Detailed Design

It is important to perform a comprehensive system review before detailed designs of the product or equipment are drawn up. Part of this process includes reviewing parameters of the manufacturing plant. The review should take into consideration parameters such as parts availability, inventories, buffer capacities and so on. Such an activity is also referred to as maintenance management. At this stage, the overall system is reviewed, not the specific equipment design parameters.

At this stage the equipment has not been designed yet. Therefore, it is not possible to precisely determine quantitative reliability criteria. It is not possible to predict system failure rates. At this stage of the design review, critical areas are identified which could potentially lead to a total plant shutdown if a breakdown were to occur. Information concerning nominal production rates, buffer capacities, operational contingencies and so on is used to perform this stage of the review.

During this stage, it should be ensured that the pertinent equipment design teams have been thoroughly apprised of the presence of any critical areas of the plant as discussed (Thompson, 1999).

5.5.3 System Review Post Equipment Design

During this stage of the design review, designers complement the initial design by considering groups of equipment that have commonalities with seemingly different groups. Examples of such groups include equipment groups that cut across

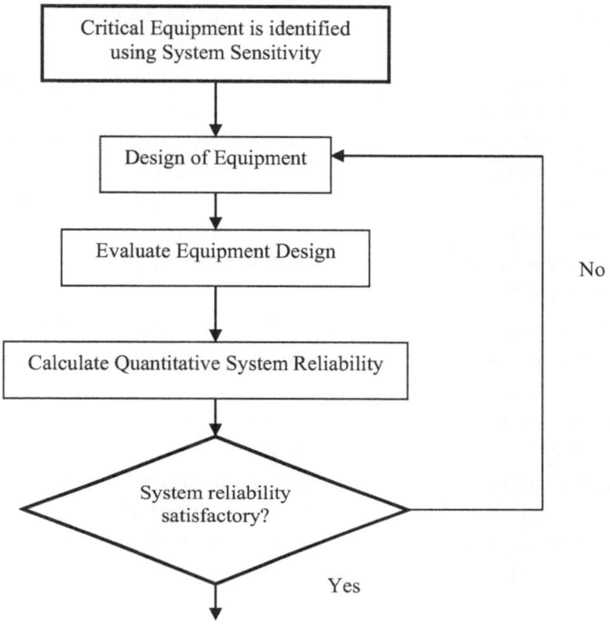

FIGURE 5.3 Design review: System and equipment design levels interaction (Adapted from Thompson, 1999).

conventional system boundaries. For instance, consider that different types of pumps in a plant are being reviewed. During this process, the designer will discover if a large number of manufacturers for such pumps exists. If so, it could lead to high requirement for spare.

One of the main principles of designing equipment for ease of maintainability states that it is advisable to define and analyze equipment groups so that uniformity is maximized. This practice helps reduce the need for spare parts. Maintenance teams can easily develop knowledge and competence in the practice of maintenance design by avoiding diverse products (Thompson, 1999). The role played by system review in the design process is depicted in Figure 5.3.

5.5.4 Evaluation of Equipment

Different items of evaluation techniques are used to evaluate different types of equipment. This stage of the design review imparts an opportunity to the design team to quantitatively evaluate a design. The following is a discussion of some of the pertinent techniques proposed by various practitioners:

5.5.4.1 Evaluating the Concept

Pugh (1991) proposed this method of evaluation. It compares potential design concepts to a reference design concept. A "reference concept" is defined as a standard design that can be considered as being barely acceptable. Any less would mean rejection.

Sometimes, a standard design could be drawn from one of the proposed concepts themselves. If upon initial inspection, the concept appears to be the best, it could be considered a reference concept. It could even be one of the proposed concepts that appear to be the best on first inspection. This practice is rarely followed though.

The different concepts are arranged on an evaluation grid. The different concepts are arranged from 1 to m. The different evaluation criteria are arranged from 1 to n. Design comprehension is facilitated by providing a small sketch of each concept. This sketch is provided on the grid.

The reference concept is considered to be the datum level. Each proposed concept is compared against it. At this juncture, there are three different possibilities.

- The proposed concept could be better than the datum. If this is the case, a score of (+) is assigned to that concept for that criterion.
- The proposed concept could be worse than the datum. If this is the case, a score of minus (−) is assigned to that concept for that criterion.
- On the basis of the comparison, no judgment can be made. If this is the case, an "s" is assigned. It is equivalent to a score of 0.

Next, the scores for each concept are added together. The best design corresponds to the one that has the highest score. Opportunities for further improving upon this chosen concept are further explored. It is still possible to modify the best design because it could still have some negative or null score corresponding to particular criteria.

The sequential process of generating ideas and concepts for design review is depicted in Figure 5.4. The figure is clear enough that it does not really need any further elaboration.

5.5.4.2 Device Performance Index (DPI)

Alternative design proposals and detailed equipment design are evaluated using DPI. Quantitative assessments are compared with respect to a range of performance parameters such as maintainability and reliability. Value judgments can also be incorporated.

The individual scores for all criteria for each design concept are combined using an inverse method. The overall value is determined by computing the DPI as given by the following formula:

$$DPI = m \times \{(1/u_1) + (1/u_2) + \ldots + (1/u_m)\}^{-1}$$

Where:

u_i are value scores for each criterion.
m = number of criteria.

Methods that simply add scores are too simple. The DPI offers several advantages over them. Consider a case where there is a low score with respect to one criterion. In that case, since the numerator is the multiple of all individual score values, its value

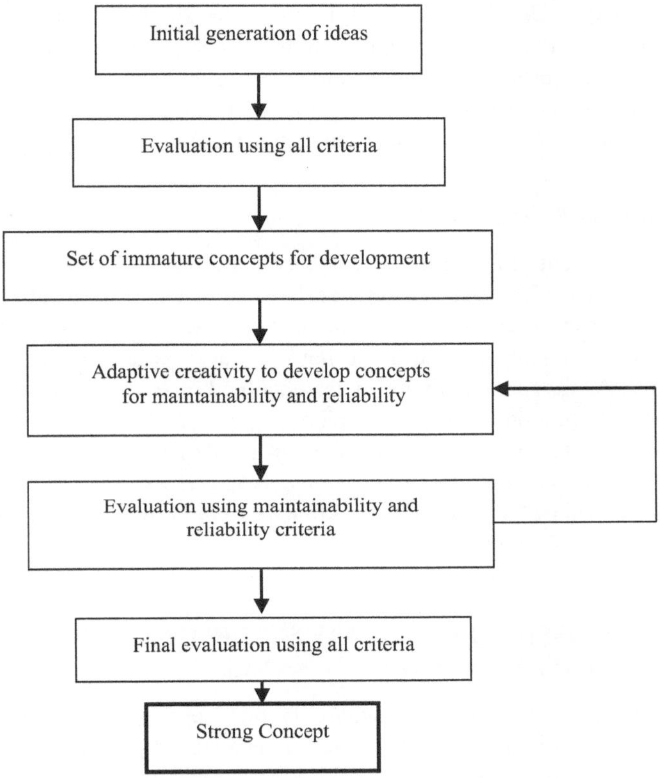

FIGURE 5.4 How do initial ideas turn into good concepts? (Adapted from Thompson, 1999).

will also be small (Thompson, 1999). Thus, the DPI of a design that scores low on maintainability will also be equally low. For this reason and because the value of DPI is directly related to the probability of success, the likelihood of that design being selected as the best design is reduced. This in turn reduces the chances of that design being selected in the final evaluation.

5.5.4.3 Analyzing the Profile of the Parameter

Performance of equipment and characteristics of the system are evaluated by this method. According to Moss and Strutt (1993), design reviews of expensive systems containing many relatively inexpensive items are often only superficial. They proposed a method to address this problem. Weak points in the system are identified by this method. Critical points where system performance has reached its limit are also identified by this method. A matrix is used to describe performance parameters that define the system with respect to the items of equipment. The system fails upon an operating performance requirement moving beyond the performance limit of an item of equipment. Consider the example of operating pressure exceeding the pressure limit specified for a valve. Using this method, it is possible to obtain a set of data points for each item. These data points pertain to relevant performance parameters for that item. All items of the product/equipment are included in the maintainability analysis. It is included in the matrix of data points. Thomson (1999) uses the "mean corrective repair time" of an item to measure maintenance performance.

5.5.5 Analysis of Components

Evaluating components of a product/equipment is quite different from evaluating an equipment/system. The main difference between a component and a system is that components can be construed as being constituent parts of the overall system. This is a question of scale. A bearing, a motor, a gasket or a rivet are all examples of components. As far as maintainability is concerned, it is impractical to consider a general survey of components in a manufacturing facility. It is important to identify certain classes of components so that detailed analysis may be performed. It is easier for experienced personnel to classify components into individual classes as detailed here. Components classified based on functionality could include seals used in fluid containers and welded joints.

5.6 MODELS USED TO FACILITATE EQUIPMENT DESIGN FOR EASE OF PREDICTIVE MAINTENANCE

Using prediction procedures for maintainability, the designer can forecast the impact specific design features are likely to have on repairing the product. It also plays an important role in determining if any additional maintainability features may be required (Harring and Greenman, 1965). It is possible to predict beforehand the down-time that can be expected to maintain the product even before it is commissioned for use. In addition, product features most likely to cause trouble during operation can be accurately pinpointed.

In this section, we will present a few design methodologies that are used to design products/equipment for ease of maintenance. Mathematical techniques that deal with equipment scheduling for maintenance are beyond the scope of this book and for that reason are not included in this chapter.

5.6.1 The RCA Method

Support time is used as a criterion for determining maintainability using the RCA method. The method makes extensive use of a checklist of physical features of product design. The following attributes directly influence computation of support time.

- Features pertaining to physical design of the product.
- Support requirements.
- Maintenance crew required to perform maintenance activities.

The physical aspects of a product are based on individual design features. Design features also help to determine the impact that layout, accessibility and packaging

TABLE 5.2
Checklist A for RCA method: Physical design features—Partial representation

No.	Physical design features	Score
1	*Access*	
I	Access adequate for both visual and manipulative tasks (electrical and mechanical)	4
II	Access adequate for visual, but not for manipulative tasks	2
III	Access inadequate for visual and manipulative tasks	0
2	*Latches, fasteners and connectors*	
I	External latches, fasteners and connectors are captive, need no special tools and require only a partial turn for release	4
II	External latches, fasteners and connectors meet one of the above three criteria	2
III	External latches, fasteners and connectors meet none of the above three criteria	0

Source: Adapted from Harring and Greenman (1965).

have on support time. A basic set of 15 questions is used to evaluate the physical design of a product. A value is assigned to each question. This value quantifies the impact of physical time on repair time.

Using empirical data consisting of more than a hundred support incidents that have occurred during the operation of ground electronic equipment, a regression analysis is conducted. Using results of the regression, a linear equation for support time is developed. This equation is presented below:

$$Z = 3.54651 - 0.02512A - 0.03055B - 0.01093C$$

Where:

A, B and C represent measures of the three parameters.

As we can see from the equation, three parameters play an important role in determining support time. These parameters are impacted by a group of design features. A partial representation of such features is presented in Table 5.2.

It is obvious from Table 5.2 that a higher score achieved on any of the individual design features translates into better maintainability. However, the scoring system is not time based. Thus, any attempt to arrive at a meaningful metric for maintainability requires the designer to perform complex regression analysis. Also, it is not quite clear as to what is meant by the term "adequate access." How much access can be construed as being "adequate?"

Maintenance is primarily a manual activity. For this reason, the definition of accessibility is subjective. What may be enough for one person may be insufficient for another person. For instance, the girth of a heavier worker is likely to be more than that of his/her lighter counterpart. So how would you design the access to take factors such as these into consideration?

Also, the scoring system is based on empirical data, which may not necessarily have a scientific basis. These factors need to be taken into consideration when using the RCA method.

A scoring system is developed for support items, which are dictated by system design. A set of seven questions are asked and each is assigned a numeric value. This is depicted in Table 5.3.

Table 5.3 indicates a list of scores for tasks that can be performed with the help of external test equipment. A task can be completed in less time if it does not require any external test equipment. Such a task is the function of design parameters that lend themselves to ease of maintainability. The scoring system is similar to the one developed in Table 5.2 in that it is built on a base number of 4.

As the level of difficulty for each task increases, the task receives a numeric score that is equal to half that of the previous task. Checklist B does not present an exhaustive listing of key equipment. The equipment is not sufficiently described.

TABLE 5.3
Design dictating facilities: Partial representation on checklist B for RCA method

No.	Design dictates—facilities	Score
1	*External test equipment*	
I	Task accomplishment does not require the use of external test equipment	4
II	One item of test equipment is needed	2
III	Two or three items of test equipment are needed	1
IV	Four or more items are required	0
2	*Assistance (technical personnel)*	
I	Task requires only one technician for completion	4
II	Two technicians are required	2
III	More than two technicians are required	0

Source: Adapted from Harring and Greenman (1965).

TABLE 5.4
Representation of maintenance skills: Checklist C for RCA method

No.	Description	Score
1	Arm, leg and back strength	
2	Endurance and energy	
3	Eye-hand coordination, manual dexterity and neatness	
4	Visual activity	
5	Logical analysis	
6	Memory—things and ideas	
7	Planning capability and resourcefulness	
8	Alertness, cautiousness and accuracy	
9	Concentration, persistence and patience	
10	Initiative and incisiveness	

Source: Adapted from Harring and Greenman (1965).

A level of detail insofar as the description is concerned is lacking. Also, it is not quite clear as to what is meant by clauses such as "some assistance needed" or "considerable assistance needed." The subsection dealing with "assistance" could do with a greater degree of objectivity with respect to the physical ability of a healthy maintenance worker working under normal working conditions.

Checklist C consists of a series of support personnel requirements. These are imposed by system design and are presented in Table 5.4 . A score of 0–4 is assigned to each.

Each of the elements featured in checklist C is required to be assigned a score ranging from 0 to 4. These scores are assigned by moderators or supervisors or by workers who perform the maintenance activity. This makes the score subjective.

Next, we shall consider another method that is used to design equipment for maintainability, namely the federal electric method.

5.6.2 The Federal Electric Method

Complex maintenance tasks are analyzed using the federal electric method. The method also applies time analysis to help determine the ease with which equipment can be maintained. The following steps are sequentially used to apply this method:

1 Principal parts are identified.
2 The failure rate of each part is determined.
3 Time required to maintain each part is determined.
4 The information obtained in the first three steps is used to compute the maintenance time expected for maintenance.

It should be noted that the first two steps are used to achieve routine maintenance of the product. Maintenance time required for each component of the product is computed by adhering to the following seven steps whereby broken equipment is restored to working condition.

- *Localization:* During the first step, the location of the malfunction is accurately determined. This is achieved without using any auxiliary test equipment.
- *Isolation:* This step consists of using appropriate auxiliary test equipment, built-in test points and so on.
- *Disassembly:* In order to remove defective parts from a machine, full or partial disassembly is required. This is influenced by factors such as accessibility and ease of component removal.
- *Interchange:* During this step, a sound part in working condition is substituted in place of one or more parts that have failed.
- *Reassembly:* The equipment is restored to its original working condition by reassembling the components that were removed during partial or complete disassembly. In other words, the machine is put back together.
- *Alignment:* The process of alignment consists of various steps such as making adjustments, calibrations and other checks and changes. These steps need to be

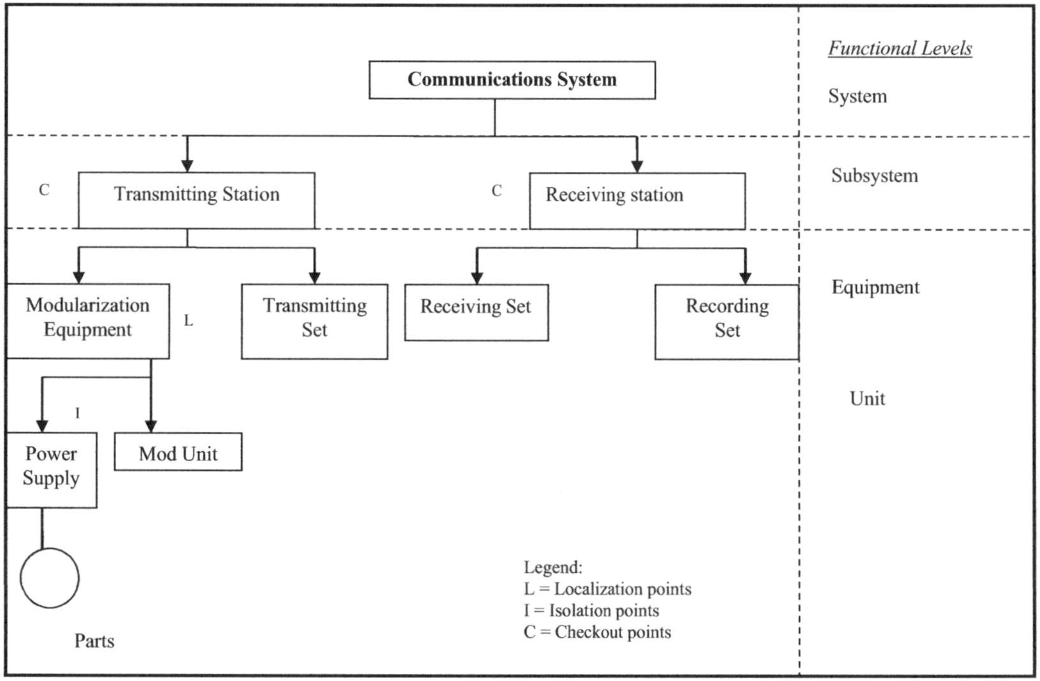

FIGURE 5.5 Federal electric method: Truncated functional level diagram of a communications system (Adapted from Harring and Greenman, 1965).

carried out on account of the repair action. In other words, the machine is fully restored to its working condition.

- *Checkout:* This is the last step. The desired level of performance is verified during this step. Maintenance crew can determine whether the equipment/machine has indeed been restored to its initial condition.

On the basis of more than 300 repair tasks, standard repair-time charts were developed for each of these actions. It is necessary to perform a functional level analysis before times can be predicted. The functional analysis is performed by analyzing the equipment. In order to perform the analysis, the equipment is broken down into a hierarchical arrangement of functional levels. This arrangement is in order of complexity (part, subassembly, assembly, unit, group and equipment). Next, a functional diagram for the different levels of complexity is developed. A truncated example is shown in Figure 5.5. This procedure enables obtainment of sharper estimates of repair times. Consider, for instance, that a higher level of repair directly results in higher disassembly, reassembly, alignment and checkout times and vice versa.

Next, some of the more commonly used design attributes are defined and their importance is described in detail. Readers should thoroughly familiarize themselves with these terminologies. They are commonly used in the field of designing for maintenance. It will also help them better grasp the concepts that are presented in the SAE method and Bretby maintainability index, both of which are described later.

5.6.3 Design Attributes for Enhancing Maintainability of a Product

Many design rules exist that can help designers in their quest for facilitating maintainability of products and equipment. But for any variety of reasons, they are often ignored during the design stage of a product. One of the drawbacks of these rules is that although they are very useful, they are nevertheless presented piecemeal and not in the form of a comprehensive methodology. But given their obvious importance in improving the effectiveness and efficiency of a design, they are being mentioned in this section. An attempt was made by Knezivic (1997) to provide an overview of design principles that facilitate designing for maintenance.

1 **Accessibility**: The primary aim of conducting an accessibility analysis is to determine the arrangement of each component and connection from the point of view of facilitating its inspection, overhaul or replacement. Other practical issues such as encumbrances of the equipment necessary to perform the task should also be taken into account (DiGironimo et al., 2004). In order to improve accessibility, it is advisable that equipment and subassemblies that need to be routinely inspected are located in such a position that they can be accessed readily and easily. Special provision for certain parts should be made in the assembly. It should be possible to connect these parts rapidly for all mechanical, air, electric and electronic connections. Consider the example of the TGV train in France. The roof panels of the train can be rapidly dismounted. Also, lateral access panels and several inspection points are provided that allow for progressive inspections to be performed quickly.

Furthermore, the location of the auxiliary equipment in the power cars and passenger cars allow maintenance crew to perform maintenance operations in an ergonomically friendly manner (which implies that it is not necessary for maintenance crew to strain themselves through adoption of unnatural postures etc.).

A maintenance-friendly product should not require other items to be removed in the course of accessing items that need to be maintained. The product design should make it easy to replace or top-up items such as lubricants without having to perform any type of disassembly (Knezevic, 1997).

2 **Modular product design**: Modular product design facilitates part replacement. In a modular product, parts that are functionally similar are grouped together into subassemblies. The subassemblies are then assembled to form the final product. In order for modularization to be truly effective, standard interface equipment such as standard couplings, joints, fits and so on is used. One of the outstanding features of modularity is that once the modules have been assembled, they do not require any further readjustments. The SAAB Gripen's (aircraft) RM 12 engine is an example of modular product design. The modular design of the engine enables easy and quick inspection. The repair process generally involves replacement of a faulty module. Accessing individual components is not required during the repair process (Knezevic, 1997).

3 **Simplicity**: It is a matter of common sense that a simpler design is inherently easier to maintain. Simplicity can be achieved by undertaking measures such as reducing the number of different parts or by reducing the part variety. It is a surprising yet true fact that no tools are required to open and close the service panels on the SAAB Gripen aircraft. Here is a case of an exceedingly simple design that has perfected the disassembly process. All control lights and switches needed during the turnaround time are positioned in the same area. These are placed together for connections for communications with the pilot for refueling.

4 **Standardization**: It is typically a good idea to use standard fasteners in product design. Standard fasteners, connectors, test equipment and materials offer several advantages. Faulty components can be easily replaced if standard sizes, materials and so on are used. Standardization also assures a basic level of quality. This avoids any unforeseen surprises later during the design process. Because standardized components are manufactured on a very large scale, they are less expensive. This is another advantage of standardization.

5 **Fool proofing**: Fool proofing implies designing and arranging components so that they can fit only in one singular configuration. This is especially important in the case of items that appear to be similar but can only be used in one application. Such items should be designed so that they don't fit into the wrong assembly (Knezevic, 1997). Incorrect assembly should be readily apparent. It should be detected during the manufacturing process itself. The

following measures help achieve effective fool proofing of components (Knezevic, 1997):

- Spacing of fasteners should be staggered if three or more fasteners are required to restrain an assembly.
- Shafts that are not symmetrical about all axes should be designed such that it is not possible to fit them incorrectly either end to end or rotationally.
- Product design should ensure that if shafts of similar lengths are used, they cannot be used interchangeably. This can be ensured by using different diameters on them.
- Two or more pipe fittings close together with the same end diameters and fittings should be avoided.
- If a particular orientation is required for flat plates, this should be clearly marked on their top and bottom faces.
- Springs of different rates or lengths within the same unit also should have different diameters.

6 **Inspect ability**: Inspect ability of a product implies that it should be easy to perform a full, nondestructive, functional check on it. Consider the example of a fuse that can only be destructively tested for the testing to be effective. Maintainable design lends itself to the ease with which important dimensions, joints, seals, surface finishes and other nonfunctional attributes can be inspected. The terms “inspect ability” and “testability” are often used interchangeably.

In the next section of the chapter, we present two methodologies that utilize these concepts. These methodologies are the SAE maintainability standard and the Bretby maintainability index.

5.6.4 The SAE Maintainability Standard

A design for maintenance was formulated by the Society of Automotive Engineers (SAE). It is recommended that this standard be utilized early during the design stage of a new product/system/machine (SAE J817-1, 1976). A hierarchy of product effectiveness was established, and serviceability, maintainability, reparability and diagnostics were defined in the report. The hierarchy of product effectiveness as defined by the SAE standard is depicted in Figure 5.6.

The SAE standard establishes a numerical value to rate an existing machine or a new conceptual machine. The machine is being rated for ease of maintenance. The SAE index assigns point values to lubrication and maintenance items. These values are based on certain requirements. Preset conditions such as location, accessibility, operation and other miscellaneous factors enable effective maintenance and lubrication.

Conventional task analysis procedures are used to describe and evaluate each maintenance operation in detail. Each suboperation that requires the use of features such as location, accessibility and so on is noted. A preconceived system of scoring is used to assign scores. A higher score represents lower maintainability and vice versa.

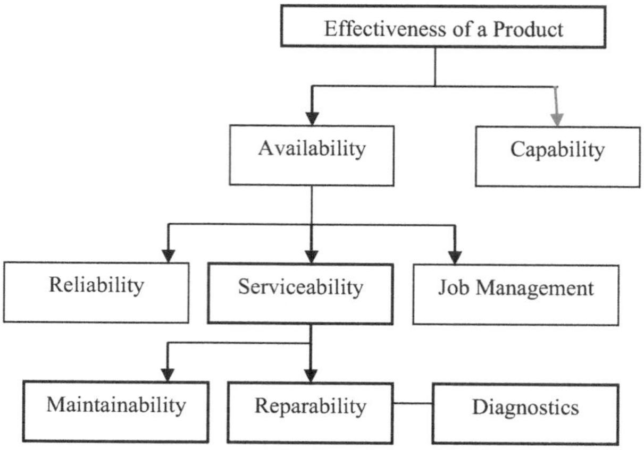

FIGURE 5.6 Hierarchy of product effectiveness (Adapted from SAE Information Report, SAE J817).

The tables that follow present briefly each requirement enumerated in the original standard. The degree of facility with which a requirement can be accomplished translates to a higher or lower score assignment.

5.6.5 Location (SAE J817)

In order to perform a maintenance task, it is necessary for maintenance personnel to adopt a certain position or posture. This is referred to by the term "location" in the SAE standard. The assumption in this section of the index is that only one operator is required. If many operations can be performed from the current position, the first operation is assigned the points applicable to that location. Each subsequent operation is assigned one point. The numeric scores attributed to the design feature "location" are depicted in Table 5.5.

It can be observed from Table 5.5 that tasks that require substantial moving around and special accessories such as ladders take longer to perform. This is also quite intuitive. For this reason, a higher score is assigned to such activities. Also, tasks that require the maintenance crew to bend, kneel and so on are difficult to perform. Again, such tasks take longer to perform leading to a higher score on the index.

5.6.6 Access (SAE J817)

The ease with which a lubrication or maintenance point can be reached is described by the term "access." Just like location, if multiple operations can be carried out using the same access facilities, the first operation is assigned points applicable to the access. One point is assigned to each subsequent operation. Accessibility point values for various design features are presented in Table 5.6.

The ease with which a maintenance point can be reached is defined as accessibility. We have already dealt with this definition in the previous section of this

TABLE 5.5
SAE index: Locations and respective point values

No.	Positions	Points
1	Ground level, working within normal reach	1
2	Ground level, bending or stretching outside normal reach	2
3	Ground level, squatting, kneeling or lying (except under the machine)	3
4	Mount machine—normal reach	10
5	Mount machine—bending, stretching or squatting	15
6	Any position (other than upright) under or within the confines of the machine	25
7	Requires climbing into position without handrails and/or steps/platforms being provided	50

Source: Adapted from SAE Information Report (SAE J817).

TABLE 5.6
SAE index: Accessibility design parameters and respective point scores

No.	Accessibility parameters	Points
1	Exposed	1
2	Exposed—through opening	2
3	Flip up cover or flap	3
4	Door or cover—hand operated	4
5	Door or cover—single fastener	10
6	Door or cover—multiple fasteners	15
7	Hood removal	35
8	Multiple covers—multiple fasteners	50
9	Radiator guard removal	50
10	Tilt cab	75
11	Crankcase/drive train guard removal—hinged and bolted	75
12	Crankcase/drive train guard removal— bolted only	100

Source: Adapted from SAE Information Report (SAE J817).

chapter. A maintenance point that can be reached easily is said to be easily accessible. It takes less time to access it and for this reason it gets the lowest score. On the contrary, a point that is exposed but is flanked by an opening gets a higher score. Constraints are imposed by the opening, which increase the amount of time necessary to access the maintenance point. If constraints are imposed by design features on the accessibility of a maintenance point, it takes more time to gain access to such a point. This results in a higher score being assigned to it. The removal of a hood gets 15 points. This is because of several design parameters. First, the hood is itself an obstruction. Second, it may be heavy and unwieldy. As such, it may take more time to remove it.

But it is not necessary that all access points are exposed. Functionality sometimes plays a role such as the one featured in Table 5.6. In such situations, a design compromise needs to be reached between functionality and accessibility of the component.

5.6.7 Operation (SAE J817)

The actual servicing or maintenance of the items in question is referred to by the term "operation." The SAE index deals with the maintenance requirements of heavy machinery such as off-road work machines. As we have already seen, the main maintenance operation can be facilitated or impeded by location and access. Table 5.7 presents the various operation categories along with the point scores assigned to each.

An abridged version of the operations section of the SAE index is presented in Table 5.7. A task that can be easily carried out gets a lower score and vice versa. Design features that facilitate the performance of such a task are scored accordingly. For instance, a visual check of a liquid compartment is easier and takes less time. Compare this to a visual check of a dipstick or one in which it is required to use a multiple screw cap and an unfastening tool. The latter are more time consuming.

TABLE 5.7
SAE index: Operation considerations and point scores (abridged version)

No.	Operation considerations	Points
1	*Compartment checking (Liquid)*	
	Visual check	1
	Dipstick	3
	Screw cap: Hand removable	4
	Multiple screw cap: Hand removable	6
	Screw cap or plug requiring tool	8
	Multiple screw cap or plug requiring tool	10
2	*Component checking*	
	Visual check	1
	Hand check of belt tension	2
	Nonprecision tool (includes tire pressure check or torque wrench)	5
	Precision tool	10
3	*Draining*	
	Drain valve—hand operable	1
	Drain valve—tool required	3
	Horizontal plug	6
	Vertical plug	8
	Cover plate	10
	Multiple plugs or covers	15

Source: Adapted from SAE Information Report (SAE J817).

TABLE 5.8
SAE index: Miscellaneous considerations (abridged version)

No.	Miscellaneous consideration	Points
1	Bleeding required	3
2	Priming required	3
3	Special tool required	4
4	Need for special instruction	10
5	Inadequate clearance for required operation	20
6	Operation requiring caution	100
7	Position requiring caution	100

Source: Adapted from SAE Information Report (SAE J817).

5.6.8 Miscellaneous Considerations (SAE J817)

Ideally, every maintenance operation should be very easy and straightforward. However, this is not always the case. Product functionality always supersedes every other design objective. For this reason, it is sometimes necessary to incorporate certain design features that will actually impede the maintenance operation, not facilitate it. But it cannot be helped.

The point values listed alongside each item are in fact punitive, also referred to as penalty points. An abridged version of this section is presented in Table 5.8.

In Table 5.8, special mention must be made of the inclusion of operations and positions that require caution. Such items pose a great deal of danger from the maintenance and operational perspective. As such, a score of 100 penalty points is assigned to each.

Also, as far as facility of performing a maintenance operation is concerned, it should be obvious by now that maintenance operations that can be done by hand alone are preferred over all others. An operation that requires a standard tool is acceptable. However, a penalty score of 4 points is assigned to an operation that requires a special tool. A special type of skill and a greater degree of skill is required to operate such a tool. Such skills may be in short supply.

An avoidable and undesirable operation receives a higher penalty score. The operation is rendered undesirable on account of the design features of the product. The SAE index aims to simplify the product design from the point of view of maintenance.

As we have seen before, predictive maintenance needs to be repeated over time. The SAE index addresses this important fact by incorporating a frequency multiplier, which is presented in the next paragraph.

5.6.9 Frequency Multiplier (SAE J817)

One-time maintenance tasks or tasks that require less than hundred hours of maintenance work are not included in the frequency multiplier. An overview of the frequency

TABLE 5.9
SAE index: Frequency multipliers

No.	Maintenance interval	Frequency multiplier
1	1000 h—semi-annually or greater	1
2	500h—quarterly, or as required	2
3	250 h—monthly	4
4	100 h—semi-monthly	10
5	50 h—weekly	20
6	10 h—daily	50

Source: Adapted from SAE Information Report (SAE J817).

multiplier assignments to different maintenance schedules is presented in Table 5.9. The maintenance hour intervals listed in the table conform to SAE recommended practice J753. Sometimes, intervals other those shown need to be used. In such cases, the frequency multiplier of the nearest SAE interval is applied. A penalty of 2 points is added to it.

5.6.10 The Bretby Maintainability Index

The SAE index was described in detail in the previous section of this chapter. While the SAE index assesses the maintenance operation comprehensively, it could still be improved upon. The Bretby maintainability index seeks to achieve this objective. Its goal is to quantify the ease with which a product can be maintained. Highlights of the index are presented in this section.

5.6.11 Description of the Index

Mason et al. (1989) described the structure of the Bretby maintainability index in detail. In order to perform the maintenance evaluation, various maintenance tasks and procedures are considered in their entirety. Each task is directly influenced by product design attributes. The index assigns time-based value to each of these attributes. The intent is to rectify some of the anomalies that were observed in the SAE index. These anomalies are listed as follows (Mason, 1990):

- It was observed that the SAE index was limited insofar as its area of application was concerned. Also, it produced a "figure of merit" for a task, not a time estimate.
- It is necessary to prepare components and surfaces prior to carrying out a maintenance operation. The SAE index fails to take this fact into consideration. Also, it does not account for the other factors such as the weight of components and size or position of access apertures. Other related design attributes such as restricted access for tools is not considered either.

- The SAE index was deemed to be relatively simple. Its capacity for handling any task that was more complicated than the basic maintenance operation was questioned. Practitioners did not find it was adequately equipped to handle more complicated maintenance tasks.

The Bretby index addresses two types of operations:

- Operations that enable maintenance personnel to gain access to equipment.
- Maintenance operations themselves.

5.6.12 Part 1 of the Index: Accessibility

This section of the index is further divided into two subsections. The removal and replacement of hatches and covers is addressed by the first subsection. This subsection answers the question: How can the machine be accessed from the outside?

The space inside openings and apertures is dealt with by the second subsection. There are other important and practically applicable factors that affect maintainability of a product. These include surface/component preparation and manual activities such as carrying/lifting. It is important that a good maintenance methodology take these factors into account.

In order to fully consider manual activities, related factors such as energy expenditure estimates and postural difficulty need to be considered. Some of the more important design attributes covered by the "access" section of the Bretby index are presented in Table 5.10.

A cursory comparison of the SAE index and the Bretby index reveals that the Bretby index accounts for some practically important points that are not considered by the SAE index.

TABLE 5.10
Access section of the Bretby maintainability index: Part 1

Hatches and covers

No.	Description	Points score
1	Flip-up cover or flap: No fasteners	3 per cover
2	Door or cover (hand-operated fasteners)	4 per cover
3	Door or cover—single fastener (tool operated)	5 per cover
4	Door or cover—multiple fasteners (tool operated)	10 per cover
5	Lift off/lift up panel—easy to handle < 12kg	2 per cover
	12–24 kg	4 per cover
	25–35 kg	6 per cover
	> 35 kg	10 per cover

Source: Adapted from Mason (1990).

TABLE 5.11
Bretby maintainability index: Location subsection of the access section

Location No	Description	Points score
1	Ground level: Working upright, within normal reach	1
2	Ground level: Bending or squatting, outside normal reach	2
3	Ground level: Squatting, kneeling or lying (not under machine)	3
4	Mount machine: Normal reach	6
5	Mount machine: Bending, stretching or squatting	8(S)
6	On Machine: Subsequent operations within normal reach	1 each
	Subsequent operations bending/stretching	2 each
	Subsequent operations squatting/kneeling	3 each
7	Any position (other than upright) under or within confines of machine	10(S)
8	Enter driver/operator cab	3

Source: Adapted from Mason (1990).

A section that addresses component location has been added to the Bretby index. This inclusion makes the index more comprehensive. Scores are assigned to machine components based on their accessibility. Components that are most easily accessible do not require the adoption of awkward and unnatural postures. The lowest score is assigned to them. Each score is further converted into a time metric. The "location" subsection of the "Access" part of the index is presented in Table 5.11.

The need for assuming awkward postures to perform maintenance procedures has been accounted for by the Bretby index. This fact is clear from Table 5.11. This consideration is important since a prolonged need to adopt such postures can lead to the onset of musculoskeletal disorders. Observe that the least scores are received by the simplest, most natural postures. This implies that they consume less time. These are also the most ideal of all postures in the list.

Conversely, it is possible to accurately determine product design attributes that complicate the maintenance task by virtue of the need to adopt unnatural postures. The designer can focus on simplifying the design, so it is more maintenance friendly.

5.6.13 Part 2 of the Index: Operations

The operations section of the index consists of 12 sections. Sections dealing with removing or replacing a component, carrying and lifting it and preparing a component are deemed more important. Removing or replacing a component is further modified by inclusion of an additional subsection. This subsection deals with operations that do not involve complete removal of a component or fastener. Thus, actions such as slackening a fastener are addressed by this subsection.

After the maintenance operation has been performed, fasteners that were slackened need to be retightened to return the machine to working condition. The Bretby index provides a clear subclassification of the process of component/fastener removal/replacement as described. To the best of our knowledge, this subclassification is unique to the Bretby methodology. It was not covered in the SAE index. An example of the removal/replacement index is presented in Table 5.12. Table 5.13 presents the slackening/tightening index.

The process of removal/replacement of fasteners is addressed by Table 5.12. Machine components do not always need fasteners to secure them. After the fasteners have been removed, machine components are sometimes lifted or manipulated in some way. It is necessary to make allowances to account for the weight of the component. This is especially true for heavy and unwieldy components. In the Bretby index, components that are easy to handle (weighing < 12 kg) are assigned a score of 2 points per component. It is necessary to make this provision since maintenance is largely a manual activity. Handling machine components during maintenance involve lifting, moving, and refitting them. This process consumes a lot of time. It should be obvious that lighter components are easier to maintain (at least from the point of view of material handling).

TABLE 5.12
Bretby maintainability index: Removal/replacement subsection of the operations section

Removal/replacement (fastener type)

No.	Description	Points score
1	Spin on fastener	1
2	Single fastener not requiring tool	3
3	Single fastener requiring tool	4
4	Additional fasteners not requiring tool	2 each
5	Additional fasteners requiring tool	3 each

Source: Adapted from Mason (1990).

TABLE 5.13
Bretby maintainability index: Slackening/tightening section

Slackening/tightening of components/fasteners

No.	Description	Points score
	Fastener type	
1	Single fastener not requiring tool	1
2	Single fastener requiring tool	2
3	Additional fasteners	1 each
	Fastener force requirements	
4	Slackening fastener: High forces needed Requiring impact	1 (H), (S) 1-8 (S)
5	Tighten to unspecified torque	2

Source: Adapted from Mason (1990).

5.6.14 Bretby Index: Other Features

The Bretby index has some more features apart from the ones described. The more prominent of those are described here.

- **Tasks that involve carrying and lifting:** Carrying and lifting tasks are especially important when large machines with heavy unwieldy components require maintenance. This section accounts for the frequency of lifting tasks. Also, other considerations such as provision of headroom to enable satisfactory maintenance and lifting have been considered. A one-person lifting task as against a two-person task has also been accounted for. The assumption is that one person can satisfactorily perform all lifting and carrying tasks for all objects between the weight ranges of 0 kg and 35 kg. An additional allowances section is also included in the index.
- **Tasks that involve preparation:** Prior to performing the actual maintenance operation, it is often necessary to first "prepare" the surface or component. Doing so facilitates the actual maintenance procedure. An entire section dealing with such preparation tasks is included in the Bretby index. Discrete preparation tasks have been allocated specific points. Four points are allocated to a task that requires cleaning around unions and fasteners. Twenty points are allocated to the task of jacking up and chocking the machine pre maintenance. Two points are allocated to the action of donning standard protective equipment such as gloves and goggles. Standard personal protective equipment (PPE) can be donned quickly because it is donned habitually. Five points are allocated to donning nonstandard PPE because it is not donned habitually.
- **Including factors that are practically important:** Practically important factors such as component checking, lubrication, draining and so on are allocated positive points. Tool access parameters are also considered. For example, 1 2-3 flats access for spanners/allen keys is considered to be sufficient clearance and is awarded 1 point per fastener which affords this kind of clearance. The point score is inversely proportional to the amount of clearance. Other miscellaneous items such as energy output, frequency of operations, visual fatigue and so on are also considered.

5.6.15 Using the Index

In order to use the index, a list and frequency of all maintenance tasks that need to be performed is obtained. Next, based on the principles of task analysis, each task is described in sufficient detail. Experienced engineers/fitters may be able to provide this information (Mason, 1990).

5.7 DESIGN RECOMMENDATIONS FOR EASE OF MAINTENANCE OF AIR FORCE WEAPON SYSTEMS

Highly available weapon and space systems are strategically important and play a pivotal role in the military. They are critical to national defense. This role continues to increase with the passage of time. The following factors tend to reduce the availability of such systems:

- Increased complexity of the system.
- Imposition of extraordinary environmental constraints.
- Demands for extreme accuracy.

This section of the chapter will present some of the key attributes that need to be considered when designing such systems for improved maintainability. Some of the attributes that will be considered include accessibility, handles, fasteners and so on. These design guidelines were obtained from a technical report entitled "Guide to integrated system design for maintainability" reproduced by the Armed Services Technical Information Agency (Rigby et al., 1961). Readers should keep in mind that this is a highly truncated version of all the listed guidelines presented in that report.

5.7.1 Design Requirements for Accessibility

As we have repeatedly stressed in this chapter, accessibility constitutes one of the principal design attributes that directly affects the ease with which a product can be maintained. Inaccessibility often leads to ineffective maintenance. If a maintenance technician is unable to see, reach and manipulate the items on which work is to be done, they may tend to delay or omit maintenance actions. It could also result in the technician making mistakes or accidentally damage equipment.

In order to improve maintainability of equipment, access must be provided to all points, items, components and units that require or may require testing, adjusting, servicing, removal, replacement or repair. The type, size, location and shape of the access are based upon a complete understanding of the following parameters:

- Operational location, setting and the environment of the equipment.
- The frequency with which access must be obtained.
- Type of maintenance functions that must be performed through the access.
- Amount of time it is likely to take to perform each of the maintenance tasks.
- Nature of tools and accessories that must be used to perform said maintenance tasks.
- Amount of work clearance necessary to perform the maintenance tasks.
- Type of clothing the technician needs to wear. This can include PPE.
- Distance to which the technician must reach (how far the technician must reach) inside the access.
- The visual requirements of the technician in performing the maintenance task.
- What exists behind the access and the packaging of items and elements.
- The items, units and elements mounted behind the access.
- Hazards involved in or related to the use of the access.
- It is necessary to take into consideration the size, shape, weight and clearance requirements of any logical combination of human appendages, tools and so on that must enter the access. This consideration improves worker safety when performing the maintenance operation.

5.7.2 Types of Access in Order of Preference

Six types of access are considered and are presented based on order of preference:

- Exposed or uncovered equipment: If structural, environmental, operational and safety conditions permit, it is advisable to leave equipment exposed for ease of maintenance. This is especially true in the case of test and service points and maintenance displays.
- Semi-exposed equipment: Semi-exposure is achieved by:
  1 Pullout racks or drawers.
  2 Full-length doors on equipment racks or cabinets.
  3 Quick-opening hoods or covers.
  4 Easily and quickly removable dust covers and cases.
- Limited access openings that are uncovered: Uncovered openings can be used only in a limited number of cases such as when environmental control is not required and when danger to equipment or personnel is minimal.
- Covered, limited access openings: When dealing with covered accesses, they must always be evaluated in terms of types of covers and fasteners used.
- Stress doors: High-performance equipment often employs stress doors. However, if possible, they should be avoided. If required, the accessibility of stress doors can be improved by using the appropriate type of fastener.
- Riveted panels and doors: Riveted panels are never acceptable as access points. The overall layout and design of equipment should be such that it should not require the removal of permanently attached structures even for maintenance that must be performed infrequently.

5.7.3 General Design Requirements for Accessing a Component

1 When it is possible and feasible to do so, the focus of the design activity should always be to design on a "grand scale." This is achieved by adhering to the following guidelines.
   - If possible, use split-line design.
   - Hinged or removable chassis should be used.
   - Major units and assemblies such as engines, turbines and so on should be designed with removable housings. This facilitates complete inspections.
   - Since we are discussing ease of maintenance of weapon systems, aircraft or missile skin should be hinged for ease of access to assemblies and accessories during major checkouts and turnarounds.
   - The design of unit accessibility features should be correlated with accessibility requirements of the overall system.

2 Accesses should be designed, located, covered and fastened such that the necessity for removing components wires and so on to reach the item that requires maintenance is avoided.

3 The removal of replaceable units should require opening of only a singular access. This applies unless the accesses incorporate a latched or hinged door.

4 Items requiring visual inspection such as hydraulic reservoirs, gauges and so on should be located such that they can be observed without necessitating the removal of panels or any other components.
5 The edges of accesses should be lined up with internal fillets or other suitable protection especially where there is a possibility that sharp edges might injure technicians puncture hoses and so on.
6 All maintenance operations that require visual control and where hazards can be encountered within the access should be provided with visual access. It is important that the technician be provided with this access.
7 If accesses are unavoidably located over dangerous mechanical or electrical components, the access door should be designed such that when it is opened, an internal light is turned on (by the opening action of the door) and a high visibility warning label is provided on the door.
8 Accesses that lead to equipment with high voltages should be provided with safety interlocks. If for some reason, it is necessary for the equipment circuit to be "ON" during the maintenance operation, a cheater switch should be provided that will reset automatically upon the access being closed.
9 Self-sealing tank should be provided with an access of such size and location that enables access to the entire interior of the tank for the purpose of inspection, cleaning or other maintenance tasks without requiring removal of the tank.

5.7.4 Shape of Access

It is not necessary that the access adhere to any particular shape. The access should be shaped such that it facilitates easy passage of required items such as tools, body appendages and any other required implements. The following factors should be considered:

1 Dimensions of various units that must be replaced through the access.
2 Attachments, protuberances, handles and so on that are attached to such units.
3 Method of grasping units during the removal process as well as required clearances.
4 Requirements for work clearances for work within the compartment.
5 Requirements for visual control of functions that are performed with the compartment.

Like we have said before, it is not necessary that accesses adhere to a particular geometric shape. A shape, though irregular, is preferred to a regular shape if it satisfies the structural and accessibility requirements better.

5.7.5 Location of Access

Accesses should be located:

- Only on face of units that will be accessible in normal installation.
- To permit direct access and maximum convenience for job procedures.

- On the same face of the equipment as the displays, controls, points and cables under question.
- Away from high voltages or dangerous moving parts. Adequate insulation and shielding should be provided around such parts to prevent injuring personnel.
- To enable heavy units to be pulled out rather than lifted out.
- Such that the bottom edge of a limited access is not lower than 24 inches and the tope edge is no higher than 60 inches from the floor.

5.7.6 Size of Access

The size of the access is determined by the same considerations as shape of the access. In general,

- One large access is preferred to two or more small ones. However, if required by structural or other considerations, it is possible that visual and physical accesses are provided separately.
- If stress doors are employed or access covers are difficult to remove, frequently used test or service points can be accessed using a smaller access.

The following section provides details on design requirements of handles for ease of maintenance.

5.7.7 Design Requirements for Handles

The dimensions, location and positioning of handles are determined by the following factors:

- Weight of the item or unit.
- The number of personnel (or hands) required to lift or carry the item.
- Type of clothing or gloves worn by these personnel.
- Operational position of the item relative to other items and obstructions.
- Manner in which the item is to be positioned or handled.
- The overall distance over which the item should be carried.
- The frequency with which the item should be handled or carried.
- Any additional uses that could be served by the handles.

5.7.8 General Requirements When Designing Handles

1 Handles should be provided on all packages, units, components and covers that:
   - Are difficult to grasp, remove or carry.
   - Need to be frequently handled or carried.
   - Weigh over 10 pounds.
   - Possess fragile components that might be used as handholds.

2 Hoist lugs (also referred to as lifting eyes) should be provided on all items that weigh more than 150 pounds. The message “LIFT HERE” should be marked adjacent to each lug. A minimum clearance of 4 inches should be provided above the eye for convenient use.

3 Handles, lugs and other handling equipment (such as casters, push bars etc.) should be permanent parts of the unit case.
4 Molded handles should be provided when it is necessary to carry items for long periods of time. The handles prevent undue side pressure on the fingers.
5 Handles that are recessed or concealed or folding handles may be used in order to conserve space. However, it is necessary that they are accessible without tools and must remain securely folded when not in use.

Gripping efficiency is best if fingers are able to curl around the handle or edge to an angle of 120 degrees or better.

5.7.9 Location of Handles

- Single handles should be located around the center of gravity.
- Two or four handles should be located at equal intervals from the center of gravity.
- Placement of handles should not interfere with the operation of the equipment or its maintenance.
- There should be a provision of a clearance of at least 2 inches between handles and obstructions.
- If an item must be pulled from a rack, handles should be placed on the front panel.
- Handles should be located and positioned in such a way that they can be held comfortably.
- The location of the handle should enable the carried item to ride clear of the legs of the carrier.

5.7.10 Additional Use of Handles

Handles are also designed and located in order to achieve the following objectives:

- Guard against inadvertent actuation of controls.
- Protect delicate parts or instrument faces.
- Serve as locking devices so that components can be secured in place.
- Serve as protective supports or stands for components. For instance, they can be used as maintenance stands when items are inverted.

The following section provides details on design requirements of fasteners for ease of maintenance.

5.7.11 Design Requirements for Fasteners

The following factors must be considered when designing, selecting and applying fasteners:

- Stress and environmental factors that must be withstood by the fasteners.
- Workspace, tool clearance and wrenching space around the fastener.
- Types of tools required for the operation of the fastener. This is expressed as a function of the type of fastener, application and its location.

- Types and variety of fasteners that are being used elsewhere in the system or that is being used by the organization.
- How frequently will the fasteners be operated?
- The amount of time it takes to accomplish each task that involves the operation of fasteners.

5.7.12 General Fastener Requirements

- Standardization of fasteners: Storage of spare parts can be reduced. Also, danger of damaging fasteners by use of a wrong tool or fastener for a particular application can be minimized by adhering to the following guidelines.
  - i Minimize the number of types and sizes of fasteners within the system:
    1. Use only a few basic types and sizes that can be readily distinguished from each other.
    2. Use the same type and size of fastener for an application (such as all mounting bolts for a given type of item).
    3. It should be ensured that screws, bolts and nuts of different thread sizes are clearly of different physical sizes; otherwise they may be interchanged.
    4. Requirements for special or close tolerance fasteners should be avoided.
  - ii The number of differing torque requirements within the system should be minimized:
    1. Only a few basic values should be used.
    2. These values should be used to clearly differing types, sizes or coded fasteners.
    3. Plan for and provide clearance for wrenches or socket tools with variable torque settings, when precise torqueing is required.
  - iii The number of tool types and sizes required for fastener operation should be minimized:
    1. Requirements for special tools should be avoided.
    2. Select fasteners for hand operation or operation by common hand tools.
- Materials used for fasteners:
  1. Fasteners should be made of nonferrous material. This type of material is rust resistant.
  2. Corrosion-resistant steel or nickel copper alloy may be used where stress is required.
  3. Aluminum alloy should not be threaded into aluminum alloy parts.
  4. Special or close tolerance fasteners should be avoided.
  5. Tapped machine threads should possess sufficient tensile strength to withstand maximum bolt or screw torque and an additional safety margin to allow for wear.
- Mounting of fasteners:
  1. The extent to which stripped, worn or damaged fasteners can be replaced should be considered during the product design stage. Fasteners (studs) that are part of the housing should be avoided.

2 Fastener mounting holes or other tolerances should be large enough that they allow the fasteners to be "started" without perfect alignment.
3 Small bolts or screws and not rivets should be used to attach hinges, catches, latches, locks and other quick disconnect devices.
4 Bolts should be mounted with the head up. This will enable them to stay in position if the nut were to fall off.
5 Nuts and bolts, especially the ones that are frequently operated or are poorly accessible, should be mounted so they are operable by one hand or one tool by:
   i Provision of recesses to hold either the nut or the bolt.
   ii Semi-permanently attaching either the nut or the bolt.
   iii Using double nuts or terminal bolts and similar applications.
   iv Use of nut plates, gang channeling or floating nuts.

- Number of fasteners:
  1 The number of fasteners and fastener parts should be minimized.
  2 The use of hinges, catches, latches and quick disconnect fasteners should be maximized. Doing so helps reduce the number of fasteners.
  3 Use a few large fasteners instead of many small ones. The only exception to this rule is when many fasteners are necessary to maintain a fluid or airtight seal.
  4 No more than four fasteners should be used to mount a unit. One of the more common design mistakes is when too many fasteners are used when a more rigid construction is preferred.
- Location of fasteners:

Fasteners should be located so that they:

1 Are operable without prior removal of other parts or units.
2 Are operable with minimum interference from other structures.
3 Do not interfere with each other or with other components.
4 Do not constitute a hazard to any personnel, wires or hoses.
5 Are surrounded by sufficient hand or tool clearance to facilitate easy operation. Factors such as requirements for two hand manipulation, or power tools for the same purpose, breakaway or removal of stuck fasteners should be considered.
   - Coding of fasteners:
     1 All external fasteners that are manipulated over the course of regular maintenance operations should provide strong color contrast with the color of the surface on which they appear.
     2 All other external fasteners and assembly screws should be of the same color on which they appear.
     3 The heads of "special" bolts and screws should be color coded or stamp coded in order to ensure that they are handled properly and are replaced by identical fasteners.

4 Only marking which designate the size, type or torque value of the fastener should be used. Manufacturer's names and trademarks should be omitted.
5 Fastener markings should be etched or embossed to withstand exposure to chemicals, fuels, LOX, weather and other operational conditions.
6 A singular fastener marking code should be used throughout the system. This code should be predetermined and standardized. It should also conform to prevailing standard practices.

In the following section, we present a discussion related to the different types of fasteners in order of preference.

5.7.13 Types of Fasteners in Order of Preference

5.7.13.1 Quick Disconnect Devices

Such devices are fast and easy to use. They require no tools for manipulation. They can be operated with one hand and very good for securing plug components, small components and covers. But their holding power is low. Also, they cannot be used when a smooth surface is required. The general requirements for such fasteners are listed below:

Requirements for quick disconnect devices:

i These types of fasteners should be carefully evaluated based on the type of application.
ii These types of fasteners should be used wherever possible for components that must be dismantled or removed frequently.
iii The device should fasten and release easily, without the use of any tools.
iv The device should fasten or unfasten in a maximum of one complete turn.
v It should be obviously apparent when they are incorrectly engaged.
vi When many such devices are used and misconnection is possible, the female section should be located, shaped, sized or coded so that only the corresponding male section may be attached.

5.7.13.2 Latches and Catches

These types of devices are very fast and can be used easily. They require no tools, possess very good holding prowess and are especially good for large units, panels, covers and cases. Once again, it is not possible to use them when the application calls for a smooth surface. The general requirements for such fasteners are listed below:

Requirements for latches and catches:

i Long latch catches should be provided to minimize accidental springing.
ii Latches and catches should be located and positioned so they won't open accidentally under expectable operating conditions.

iii Catches should be spring loaded, so they do not require positive locking, but do so on contact.
iv A latch loop and locking action should be used if positive locking is necessary to meet structural or stress requirements.
v If a handle is used together with a latch, the latch release should be located on or near the handle. This necessitates only one hand for operating the device.

5.7.13.3 Captive Fasteners

Depending upon their type, these types of fasteners are more difficult to use. They require the use of common hand tools for manipulation. But they stay in place, save the amount of time spent handling and looking for bolts and screws and require only one-handed operation. The general requirements for such fasteners are listed below:

Requirements for captive fasteners:

i Captive fasteners should be used wherever "lost" screws, bolts or nuts might cause a malfunction or excessive maintenance time.
ii Use only those fasteners that can be operated by hand or common hand.
iii Only use fasteners that may be easily replaced in case of damage.
iv Self-locking, spring action should be provided on captive fasteners of the quarter turn type.

5.7.13.4 Combination—Head Bolts and Screws

These types of fasteners should be preferably used over other screws or bolts. The reason for this is because they can be operated more rapidly with either a wrench or a screwdriver. This allows more convenient tools to be used and the possibility of slot damage and stuck fasteners is reduced. Generally, slotted hexagonal heads are preferred to knurled and slotted heads.

5.7.13.5 Regular Screws

Round, square or flat head screws require more time and are more subject to loss, slot damage, stripping and misapplication than their combination head counterparts. But they also require less wrenching space and one handed action to operate; too many extra parts are not involved.

Square head screws are generally preferred to round or flat since they provide better tool contact, are less subject to slot damage and can be removed with pliers. The general requirements for such fasteners are listed below:

Requirements for regular screws:

i It should take less than 10 turns to tighten or loosen a screw.
ii Upon being tightened, the screw must fully engage to a distance of at least equal to its diameter.
iii Deep slots must be provided on screw heads to minimize slot damage.

iv Screws should be used only when screw drivers may be used in a "straight in" fashion. The use of offset screw drivers should not be required.
v A tool guide must be provided in the assembly if a screw has to be operated unseen.
vi Fine thread screws are recommended for pressurized units.
vii Countersunk screws should be used only where necessary in order to provide a smooth surface.
viii Round head screws should be used instead of slat head screws on panels that are less than 3/32 inch thick. This prevents screws from ripping through the panel.
ix Self-tapping screws should have one type of head and be of one size where feasible, or a minimum number of sizes.

5.7.13.6 Bolts and Nuts

Usually, bolts are slow and cumbersome to use. They require two-handed operation. Access needs to be provided to both ends of the bolt. The use of two tools is also often needed. They require precise movement in starting nuts and have many loose parts that should be handled and have the potential of being lost (such as nuts, washers etc.). The general requirements for such fasteners are listed below:

Requirements for bolts:

i Bolt length should not exceed what is required for a particular application.
ii Bolt threads should be no finer than what is dictated by strength requirements.
iii It should take less than 10 turns to tighten a bolt.
iv Upon being tightened, the bolt should extend a minimum of two threads beyond the nut.
v Hexagonal head bolts should be used in general, but especially for torque usages.
vi Left hand threads should be used only when stress conditions require and both bolts and nuts should be clearly identifiable by markings of shape or color.
vii Self-locking bolts in tapped holes should be used only when one face is smooth or is inaccessible and temperatures are below 250° F.

Requirements for nuts:

i Regular hexagonal nuts in a few easily distinguishable sizes should be used.
ii Different sizes of nuts should be used for different thread requirements.
iii Wing or knurled nuts require no tools. They should be used for low tension applications. Of the two, wing nuts are easier to use.
iv Self-sealing nuts should be used for fastening equipment to fluid tanks in order to prevent leaking around the fastener.
v Lock nuts may be used for mounting light components. However, they must withstand heat requirements. They cannot be used when fallen nuts could damage equipment.

vi Clinch nuts should be incapable of moving or rotating with respect to the surface on which they are mounted.
vii Floating nuts should have an allowable shift of only +/– 1/16 inch.

5.7.13.7 Internal Wrenching Screws and Bolts

These types of fasteners allow higher torque, better tool grip and less wrenching space. They are easily damaged, are difficult to remove when damaged and require special tools. The general requirements for such fasteners are listed below:

Requirements for internal wrenching screws and bolts:

i The number of different sized special tools should be minimized to require as few as possible.
ii Slots should be deep. This minimizes damage to the fasteners.
iii Design should allow and plan for the removal of damaged, internal wrenching fasteners in terms of clearances, power outlets and so on.
iv Requirements are otherwise the same as for bolts and screws.

5.7.13.8 Rivets

Rivets are used for permanent fastening. Permanent fasteners are very difficult to remove, and the process is very slow. They should not be used on any part that may require removal. As far as maintenance is concerned, processes such as wire stapling are preferred to rivets. The general requirements for rivets are listed below:

Requirements for rivets:

i Rivets should not be used on latches, hinges or retainers.
ii Rivets should be made out of softer material than the material that they fasten.
iii The heads of countersunk rivets should be larger than the thinness of the pieces they fasten. This prevents them from ripping through.
iv Shear rivets do not expand to fill the hole. Thus, holes must be drilled to close tolerances. Also, maintenance instructions should specify the tolerances and the sizes of plug gauges and reamers to be used.

Often, fasteners are used in conjunction with many accessories such as washers, nut plates and so on. This section presents some design guidelines about such accessories.

Washer requirements:

1 Washers should fit tightly against the underside of the fastener head.
2 Washers should fit the shaft snugly, but it should also be easy to remove.
3 Split-ring lock washers should be used with static loads that exceed 2 oz.
4 Lock washers should be used together with lock nuts. This enables maximum locking action.

Requirements for metallic inserts and blocks:

These should be secured such that tightening of the screw or bolt will not loosen or move the insert or block.

Gang channeling: Gang channeling of nuts results in saving time when handling many nuts when they are in a straight line. Only channels should be used in which nuts can be replaced individually.

Nut plates: These are heavy and expensive to replace when a hole is stripped. But they may be used when several bolts need to be fastened on one surface and alignment is not a problem.

Cotter key and pin requirements:

i Keys and pins should fit snugly but should not require driving in or out.
ii Heads of cotter keys should be large in order to facilitate removal and prevent the keys from slipping through.

Requirements of safety wire:

i Safety wire should be used only where self-locking fasteners or cotter pins are inadequate to withstand the expected vibration or stress.
ii Safety wire should be attached so it can be easily removed and replaced.

Requirements for retainer rings:

i Only use rings that can be removed and replaced easily when they are worn.
ii Rings that hold with a positive snap action should be used if possible.
iii Spring tension should be used to prevent loosening of twist-to-lock rings.

Retainer chains:

Retainer chains or locking bars are used to:

i Prevent hatches or doors from opening too far and springing their hinges.
ii Turn doors or covers into useful shelves for the maintenance personnel.
iii Prevent small covers, plates or caps from being misplaced.
iv Secure small, special tools to the location where they will be used.
v Secure objects which may otherwise fall and cause personnel injury.

Requirements for retainer chains:

i Only link, sash and woven-mesh type chains should be used. Bead link chains are not recommended since they break more easily than other types of chains.
ii Chains should be attached with screws or bolts. The attachment should be strong and positive but can be disconnected easily when required.
iii Eyelets should be provided at both ends of the chain for the attaching fasteners.
iv Chains should be no longer than necessary in order to fulfill their function.
v Chains attached to filler caps should be connected externally. Doing so facilitates replacement and prevents broken parts from damaging equipment.
vi If chains were to interfere with moving parts, they should be used.
vii Chain covers prevent chains from becoming tangled. If such covers are used, they should be flexible, easy to bend and durable.

5.8 CONCLUSION

Design for maintenance constitutes one of the more important design objectives as far as sustainable product/equipment design is concerned. Products that are easier to maintain are likely to last longer. It is necessary to consider maintainability of products early during the design phase by incorporating design factors such as accessibility, location and so on. Individual components and their spatial arrangement should be such that broken-down products/equipment can be "fixed" easily and with as little labor and time as possible. To that end, ergonomic considerations were addressed in this chapter. We also presented several important design guidelines and design methodologies that could prove invaluable to the practitioner.

REFERENCES

Adair-Heeley, C. (1989). "The JIT challenge for maintenance," *Production and Inventory Management Review with APICS News*, Vol. 9, No. 9, pp: 34–35.

Adams, S.K. and Patterson, P.J. (1988). "Maximum voluntary handgrip torque for circular electrical connectors," *Human Factors*, Vol. 30, No. 6, pp: 733–745.

Crawford, B.M. and Altman, J.W. (1972). "Design for maintainability," in: Vancott, H.P. and Kinkade, R.G. (eds.), *Human Engineering Guide to Equipment Design*, pp: 585–631, US Department of Defense, Washington, DC.

Di Gironimo, G., Monacellia, G. and Patalano, S. (2004). "A design methodology for maintainability of automotive components in virtual environment," International Design Conference-Design. Dubrovnik, May 18–21, 2004.

Eade, R. (1997). "The importance of predictive maintenance," Iron Age New Steel, Vol. 13, No. 9, pp: 68–72.

Ferguson, C.A., Mason, S., Collier, S.G., Golding, D., Graveling, R.A., Morris, L.A., Pethick, A.J. and Simpson, G.C. (1985). "The ergonomics of the maintenance of mining equipment (including ergonomics principles in designing for maintainability)." Final Report on CEC Contract 7247/12/008. Institute of Occupational Medicine (IOM Report TM 85/12), Edinburgh.

Gits, C. (1992). "Design of maintenance Concepts," *International Journal of Production Economics*, Vol. 24, No. 3, pp: 217–226.

Harring, M.G. and Greenman, L.R. (1965). *Maintainability Engineering*. Duke University.

Herbaty, F. (1990). *Handbook of Maintenance Management Cost Effective Practices*. Second Edition. Noyes, Park Ridge, NJ.

Hubka, V. (1994). "Scienza della progettazione," *Masson Editoriale*, pp: 23–56.

Imrhan, S.N. (1991). "Workspace design for maintenance," in: Mital, A. and Karwowski, W. (eds.), *Workspace, Equipment and Tool Design*, pp: 149–174, Elsevier, Amsterdam.

Johnson, A.D. (1988). "The design of modern road headers," in: *Colliery Guardian*.

Knezevic, J. (1997). *Systems Maintainability Analysis, Engineering and Management*. First Edition. Chapman and Hall, London, UK.

Macaulay, S. (1988). "Amazing things can happen if you… keep it clean," *Production*, Vol. 100, No.5, pp: 72–74.

Maggard, B. and Rhyne, D. (1992). "Total productive maintenance: A timely integration of production and maintenance," *Production and Inventory Management Journal*, Vol. 33, No. 4, pp: 6–10.

Mason, S. (1990). "Improving plant and machinery maintainability," *Applied Ergonomics*, Vol. 21, No. 1, pp: 15–24.

McDaniel, J.W. and Askrein, W.B. (1985). Report submitted to U.S. Air force Aerospace Medical Research laboratory. Wright Patterson AFB, Ohio.

Morgan, C.T., Cook III, J.S., Chapanis, A. and Lund, M.W. (1963). *Human Engineering Guide to Equipment Design.* McGraw Hill, New York.

Moss, T.R. and Strutt, J.E. (1993). "Data sources for reliability analysis," Proceedings of the Institution of Mechanical Engineers, Part E, *Journal of Process Mechanical Engineering*, Vol. 207, pp: 13–19.

Nakajima, S. (1989). *Total Productive Maintenance Development Program: Implementing Total Productive Maintenance.* Productivity Press, Cambridge, MA.

Oborne, D.J. (1981). *Ergonomics at Work.* John Wiley and Sons, New York.

Pugh, S. (1991). *Total Design.* Addison-Wesley.

Rigby, L., Cooper, J. and Spickard, W. (1961). "Guide to integrated system design for maintainability." Contract No. AF 33(616)-7059. Behavioral Sciences Laboratory, Airforce Systems Command, United States Airforce. Wright Patterson Airforce Base.

Reiche, H. (1994). *Maintenance Minimization for Competitive Advantage.* Amsterdam.

Reliability Centered Maintenance for Aircraft Engines and Equipment (February 8, 1985) MIL STD 1843 (USAF).

Reliability Centered Maintenance Handbook (January 1983 revised) Department of the Navy, Naval Sea Systems Command, S 9081-AB-GIB-010/MAINT.

Scarf, P.A. (1997). "On the application of mathematical models in maintenance," *European Journal of Operational Research,* Vol. 99, pp: 493–506.

Smith, A.M. and Hinchcliffe, G.R. (2004) *RCM-Gateway to World-Class Maintenance.* Elsevier Butterworth-Heinemann, UK.

Smith, R.L., Westland, R.A. and Crawford, B.M. (1970). "The status of maintainability models: A critical review," *Human Factors*, Vol. 12, pp: 271–282.

Swanson, L. (2001). "Linking maintenance strategies to performance," *International Journal of Production Economics*, Vol. 70, pp: 237–244.

Thompson, G. (1999). *Improving Maintainability and Reliability through Design.* Professional Engineering, London and Bury St. Edmunds, UK.

Tichauer, E.R. (1966). *The Biomechanical Basis of Ergonomics.* John Wiley and Sons, New York.

Van Cott, H.P. and Kinkade, R.G. (1972). *Human Engineering Guide to Equipment Design.* US Department of Defense, Washington, DC.

6 Consideration of Reuse, Recycling and Remanufacturing

6.1 INTRODUCTION

The traditional manufacturing process is referred to as linear manufacturing or cradle-to-grave manufacturing. It consists of extracting raw materials from the natural environment, processing it into finished products and disposing of it at the end of its useful life. They are disposed of in landfills. We have seen repeatedly throughout the course of this book that this practice is inherently not sustainable. It is essential to close the "loop" on the manufacturing process. Doing so would transform the system of linear manufacturing into cyclical manufacturing.

Cyclical manufacturing is also referred to as the cradle-to-cradle approach to manufacturing. In this approach, the material that was introduced into the system at the beginning never leaves that system. It stays in it because it is reused, recycled or remanufactured. Reuse, recycling and remanufacturing constitute the so called end-of-life (EOL) options. They answer the question: "What is to be done with all these products when they reach the end of their lives?" If the goal is to conserve the environment and prevent the landfills from being overexploited, then the only feasible alternative is to reuse, recycle or remanufacture end-of-life products. Figure 6.1 demonstrates how the loop can be closed on material flows.

However, this is easier said than done. It is not so simple to recycle a product or to reuse or remanufacture it. But aren't old cars shredded all the time? Shredding a car is not the same as recycling it intelligently. In order to truly conserve material and energy, it is essential to design the product for a particular end-of-life option. This is done by taking specific actions during the design stage itself. A designer can truly embrace the idea of sustainable product design and development only by making sure that a product is "destined" for one of the three end-of-life options. What does this mean? It means that the product has been designed in such a manner that when its time comes, it is easy to be either recycled, reused or remanufactured. Thus, the design of such a product would be substantially different from its counterpart that is destined for the landfill. The difference can lie in the choice of material, types of fasteners and joints, product structure and so on.

The goal of this chapter is to introduce the reader to design decisions that must be made early in the product lifecycle (particularly in the design stage) to make an environment-friendly product.

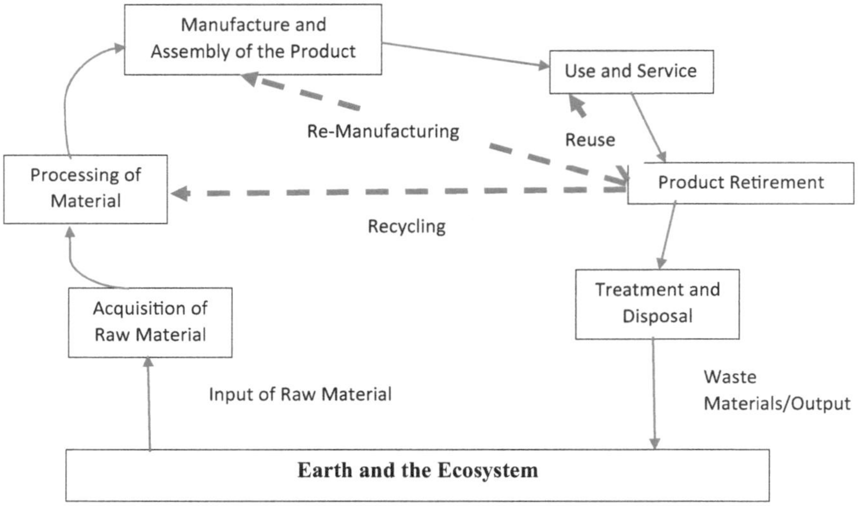

FIGURE 6.1 Closed loop material flow (Adapted from Nasr and Thurston, 2006).

6.2 WHAT IS PRODUCT REUSE?

Reusing a product implies using the product again for the same purpose or for a different purpose at the end of its useful life. This is clearly distinguishable from the concept of recycling wherein the product is broken down to its most basic material form.

If a product is destined to be reused at the end of its useful life, it is necessary to adopt proper quality assurance measures of used parts (Kimura et al., 1998). The reliability of the product throughout the entire life cycle should also be checked (Hata et al., 2000).

From the design point of view, a product that is destined to be reused must be designed and manufactured differently from other products. A different idea about the life cycle needs to be adopted in order to design products for ease of reuse. This section of the chapter will present some of the design principles that facilitate product reuse.

Many different types of products are routinely designed for reuse. For instance, in the case of single-use cameras, mechanical units are completely reused many times over and across successive generations of products (Tanaka, 2000). In the case of photocopiers, on the other hand, comprehensive systems are designed to facilitate product take back. Also design concepts such as product modularization and standardization of parts are used when designing the photocopier.

6.3 PRODUCT MODULARIZATION FOR EASE OF REUSE

Modularization is a design principle commonly used in product design to facilitate reuse. In its most basic form, a module can be defined as a building block of any product structure. Modular design has been used for thousands of years. In more recent times, it is necessary for a module to possess something more than merely being a building block. It should also possess a considerable amount of functionality. In an industrial context, such functionality should be strong enough that it can be tested independently. For instance, the power supply module in a printer has enough functionality and it can also be considered a building block of the printer (Miller and Elgard, 1998). We have already examined the role played by product functionality in chapter 1 of the book.

Traditionally, whether to adopt a modular product structure has been determined by several factors as follows (Kimura et al., 2001):

- Commonality among products, also referred to as standardization.
- Functional independence.
- Product cost.
- Ease with which product can be manufactured.
- Ease with which product can be maintained.

It is obvious that some of the above factors can have conflicting effects of the actual design of the module. Thus, it is necessary to make design tradeoffs based on their relative importance. However, since the goal of this section of the chapter is to examine product modularization to enhance ease of reuse, we will focus on that objective.

In order to enhance the possibility of reuse, the following modular characteristics are important:

- Stability of technology.
- Ability to upgrade functionality.
- Longevity of the part.
- Ease with which quality assurance can be achieved.
- Ease with which the part can be cleaned, repaired and so on.

These considerations are at odds with traditional thinking in terms of modularizing parts. The modularizing procedure with the sole aim of enhancing product reusability is delineated stepwise as follows (Kimura et al., 2001).

1 Functional dependency: During this step, the modules and/or parts of the products being designed are described by means of a graph structure. The nodes of the graphs are modules/parts and the functional or other relationships are depicted by means of arcs. The design requirements dictate the level of detail of the graph structure. During the early design stage, only functionally important components are identified. The graphs are then superimposed with identification of same or similar nodes as those that existed in part products with similar functionality. Multiplicity of superimposition with respect to each arc is counted for the next step.
2 Commonality: During this step, the number of multiplicity of arcs is identified. The number of multiplicities is referred to as weight. If the difference in values of arc weight is within a specified parameter value, say "L," then the

nodes that are connected by those weighted arcs are combined together in the form of a module. If the parameter L has different values, it leads to different module structures. A larger value of L implies less importance to structural commonality within modules.

3 Module types: In terms of types of modules, two different types can be identified as follows:
   - Component swapping modules: In this type of modules, several different exchangeable (swappable) modules are connected with a component to create a variety of products.
   - Component sharing modules: In this type of modules, various modules that share the same basic component are responsible for creating different variants of the product.

   Swappable modules and/or shared modules are combined to create new modules.

4 Dividing the modules: Modules that have similar property are combined by dividing different portions from the main parts. This is done for pairs of parts/modules that exhibit component sharing modularity. Such an operation results in decreasing the total number of modules. The aforementioned steps are repeated in order to achieve a stable product structure.

Although reusing parts has several advantages, there are some disadvantages associated with the process. For instance, rapid technological advancement can render old parts obsolete whereby they can no longer be reused. It could also be quite difficult to collect, clean and refurbish old parts. This could make it prohibitively expensive to reuse old parts.

6.4 WHAT IS RECYCLING?

The Environmental Protection Agency (EPA) defines recycling as "the process of collecting and processing materials that would otherwise be discarded as trash." Instead these materials are turned into new products by further processing the underlying raw material. According to the Europe 2020 strategy, "the quality of our life and the functioning of the global economy is underpinned by natural resources." This implies that human civilization is entirely dependent on natural resources for sustenance. Thus, it is essential to obtain reliable access to critical raw materials as well as make smarter use of natural resources. Recycling aims to close the loop of materials and components. It accomplishes this by reusing/utilizing them for new products.

Recycling, as a rule, is performed at the material level. It is rarely, if at all, beneficial to recycle the entire product. The main goal of recycling is to maximize the amount of recycled resources while investing the least amount of effort (Kriwet et al. 1995). We will examine the minute differences between terms such as recycling, remanufacturing, refurbishing and so on later in the chapter, specifically when we discuss the concept of product remanufacturing. In this section of the chapter, we will take a deeper look into the concept of design for recycling and pertinent rules and principles.

6.5 TYPES OF RECYCLING

Recycling activities can occur in three distinct loops in a product lifecycle (Kriwet et al., 1995).Two more concepts related to recycling plastic are also discussed They are described as follows.

1. Recycling of production scrap: During this process, residues emanating from traditional manufacturing processes such as punching, injection molding and so on are recycled at the material level so as to be used in new products.
2. Recycling during product usage: This involves continuing to use the product after it has been reused or remanufactured.
3. Recycling after product has been used: This involves the process when a product is recycled at the end of its useful life. Different recovery options are available at the end of life of most consumer products such as household appliances, automobiles and so on. These products have a complex product structure and composition. In such cases, other product recovery processes must take place prior to the actual process of recycling, namely disassembly, shredding and separation.
4. Closed loop recycling for plastics: Plastics constitute some of the most commonly recycled material. In closed loop recycling, the recycled plastics are used to produce the same product from which they were recovered. The new product could contain only recycled plastic, or it could be a combination of virgin plastic and recycled plastics (Ragaert et al., 2017). In the latter case, the product can continue to be recycled in the future and the material recovered from it can be added at the same rate. This is very commonly done in the case of PET (Polyethylene terephthalate).
5. Open loop recycling for plastics: In the case of open loop recycling, the recycled plastic is used in a different product. This means that the new product is unlike the old product from which the plastic was originally recovered. The new application is not necessarily of lower value. Manufacture of textile fibers from bottle (PET) or printer components from water bottle carbonates are examples of open loop recycling (Kunststofindustrie, 2015).
6. Steps in the mechanical recycling process: The process of mechanical recycling of plastics consists of the following steps, each of which can occur anywhere between not at all and multiple times in the sequence (Ragaert, 2017):
   - Separation and sorting: This step occurs based on shape, density, chemical composition, size or color.
   - Bailing: On the basis of physical location where the plastic is sorted, bailing takes place. If the processing occurs at a different location than the sorting, the plastic is bailed in between to facilitate transport.
   - Washing: The plastic is washed to remove organic contaminants.
   - Grinding: This step reduces the size of the plastic product to flakes.
   - Compounding and pelletizing: This is an optional step. During this step, the flakes are reprocessed into a granulate. The granulates are easier to use in converters.

As far as product disassembly is concerned, we have already examined this process in detail in a previous chapter of this book. But insofar as recycling is concerned, disassembly enables decomposition of the product into subassemblies or individual components, which retain their original form and characteristics.

Shredding and separation are processes that are characterized by decomposition of components and parts into random, minute pieces which can then be separated using a variety of different methods, mostly in accordance with individual material properties.

6.6 COMPONENTS OF A RECYCLING SYSTEM

The recycling system can be construed to comprise designers, recyclers, suppliers and consumers. We have already discussed the role of the designer and examined the overall product design process in the first chapter of this book. The role of recyclers to that process needs to be added.

Recyclers perform the function of dismantling the product and provide waste management services. This function can be provided by original equipment manufacturers (OEMs), an independent recycler or a contract recycler (very similar to the types of operators in remanufacturing). In this overall system, the designer controls the network and provides recycling data and assesses different decisions.

Different types of data interchange occur between the entities named above within the overall recycling system that are relevant to the end-of-life stage. The following information exchange occurs between each pair of entities.

1 Between designer and recycler:
   - Data concerning different modes of collecting, transporting ad storing the products at the end of their useful life.
   - Data concerning availability of different recycling methods.
   - Information pertaining to different markets for materials.
   - Information on properties of different materials in order to facilitate ease of sorting and separation.
   - Information concerning the existence of reusable components and their specific location, materials, whether they are harmful or valuable and information regarding how best to disassemble the product.

2 Between designer and consumer:
   - This information exchange involves tracking the exchange of parts of the product during the repair or upgrading process. The objective of this exchange is to reduce the uncertainties inherent in product recovery.

3 Between designer and suppliers:
   - Information exchange concerning the use of recovered materials in new parts, the particular specifications or reuse components.
   - Information exchanges about markets for particular materials.
   - Information about properties and quality of recycled materials and their reliability.

6.7 DESIGN FOR RECYCLING

When making a decision regarding the end-of-life options that a product is destined for, it is up to the designer to decide this early during the product life cycle. When making such a decision, the designer needs to take into account the requirements of recycling processes that may lie 15–20 years in the future. A study of relevant literature indicates the general absence of formalized rules for designing a product for recycling. As a result, a set of guidelines have been formulated based on heuristics and are presented as follows (Kriwet et al., 1995).

6.7.1 Criteria Applicable to Individual Components

- Use of hazardous and material that is otherwise environmentally harmful should be avoided.
- Standard processes such as crushing are used to recycle materials. If a material is incompatible with standard recycling processes, its use should be avoided.
- If a material can be reutilized easily, it should be given preference over other materials that are more difficult to be reutilized.
- The product should be structured in such a way that it allows for the use of recycled components. This principle will be discussed in more detail in the section on product remanufacturing.

6.7.2 Criteria Applicable to Subassemblies

- When designing subassemblies, design for recycling is facilitated by clustering materials with similar utilization compatibility.
- As with individual components, subassemblies should be designed so that the use of recycled subassemblies is feasible.

6.7.3 Criteria Applicable to Disassembly Operations

Recycling is usually performed on individual components and rarely on entire products. For instance, the steel for an automobile body is recycled. In order to perform the recycling process, it is essential to selectively disassemble the components that are destined for material recycling. Thus, the body panels of a car need to be selectively (also referred to sometimes as nondestructive disassembly, as we have already examined in the chapter on design for disassembly) disassembled from the rest of the car prior to being recycled. Thus, the disassembly process is an integral part of the recycling operation. The criteria applicable to disassembly are as follows:

- If subassemblies consist of dissimilar materials, use of joining materials that are easy to disassemble should be incorporated.
- Joining elements that are easily accessible should be designed and used for the purpose of assembly. This is a design principle that is also important when designing products for remanufacturing.

- As we have already studied in the chapter on design for disassembly, joining elements that do not require special tools are preferred to those that require special tools.
- Joining elements that are corroded pose various problems. First, they take longer to disassemble because they need to be cleaned thoroughly prior to the actual disassembly. Second, they can pose a health hazard to the person performing the disassembly. Such joints are also more difficult to access.
- Since the material is being recycled, it is not necessary to always resort to nondestructive disassembly (unlike remanufacturing). Components can be separated using destructive disassembly as well. However, this process should be avoided if it produces sharp edges.

6.7.4 Criteria Applicable to the Entire Product

- In order to improve the recyclability of the entire product, it is preferable to minimize the overall variety of materials used in the product.
- The number and variety of joining elements used in the products should be minimized.
- If there are any harmful materials and/or components used in the product, the product needs to be designed to allow easy access to them.
- Precedence relationship between parts should be avoided in order to improve accessibility.
- Paints and laminates should be avoided where possible. The product should be built along planes to facilitate access and ease of separation.

6.7.5 Criteria Applicable to Recycling Logistics

- The product should be designed in such a way that it allows for predisassembly. This makes it easy to transport after usage.
- Information about the advantages of recycling should be widely propagated. This will occur through educating the public about them. This will encourage consumers to begin recycling the products thus initiating the recycling process.
- Information pertaining to recycling should be made easily available. This includes information on material content, disassembly procedures, recycling processes and options that are available and so on.

6.8 ASSESSING RECYCLABILITY

A study performed in 1996 assessed the recyclability of automobiles (Coulter et al., 1996). Three vehicles were disassembled and the factors that influenced their recyclability were determined. The recyclability of the automobiles was based on two ratings for the various components of the cars. Specifically, these ratings were termed as recyclability rating and material separation rating. These ratings are presented the following tables. The recyclability rating is presented in Table 6.1 and the material separability ratings is presented in Table 6.2. In Table 6.1, examples of individual parts that can be recycled are graded on a scale of 1 through 6. A score of 1 represents

TABLE 6.1
Recyclability ratings for automobiles with examples of parts

Rating	Description	Examples
1	Part is remanufacturable	Starters, alternators
2	Material in the part is recyclable with a clearly defined technology and infrastructure	Most metals Catalytic convertors
3	Technically, it is feasible to recycle the material, but the infrastructure to recycle the material is not available	Most thermoplastics Glass Seat foam
4	Technically, it is feasible to recycle the material, but further processes and material development are required	Armrests Steering wheels Airbag modules
5	Material can be used for energy recovery because it is organic but cannot be recycled	Headliners Wood products
6	Material is inorganic with no known technology that can be used for recycling	Heated glass Fiberglass headliners

Source: Coulter et al. (1996).

TABLE 6.2
Ratings for material separability

Rating	Description	Examples
1	Material may be disassembled with ease manually; it takes approximately one minute to achieve disassembly	Cover of lower steering column
2	Material may be disassembled manually, but with some effort; it takes approximately one to three minutes to achieve disassembly	Instrument cluster, radio
3	Material may be disassembled with some effort and requiring some mechanical means or shredding Mechanical means are required to separate component materials (*the process for doing so has been fully proven*)	Engines, sheet metal
4	Material may be disassembled with some effort and requires some mechanical separation and shredding; mechanical means are required to separate component materials, *but the process is currently under development*	Instrument panels
5	It is not possible to disassemble the materials; no known process exists to achieve separation	Heated backlights

Source: Coulter et al. (1996).

a part that is remanufacturable and a score of 6 represents a part that cannot be recycled at all. As far as recycling is concerned, the highest score that a part can possibly achieve according to this grading system is a "2." Examples include most metals and catalytic convertors.

6.9 MANUAL MATERIAL SEPARATION VERSUS MECHANICAL SEPARATION

It is obvious from Tables 6.1 and 6.2 that when products are designed so as to facilitate manual material separation, they are easier to recycle as compared to when material separation has to be performed mechanically. Recovery of metal from the vehicle as a whole is an example of mechanical material separation. In this case, it is possible to easily recover the steel after having shredded the vehicle. But doing so manually would be quite difficult. The distinctly different requirements of manual material separation and mechanical material separation are described here.

- Manual material separation: If the design goal is to simplify the process of material separation to the point that it can be performed manually, it requires more effort on the part of the product designer. This is because it requires more effort to improve the disassembly and sorting process. Manual separation is facilitated by reducing the amount of time required to separate and identify the material. This makes the process economically more feasible. To put it succinctly, design guidelines for this process should suggest different ways by which manual disassembly effort is reduced. Design guidelines that result in being able to more easily identify the material visually should also be included.
- Mechanical material separation: If the product is being designed with the ultimate goal wherein the materials can be easily separated mechanically, the assembly or component should be designed such that they can be separated easily and quickly into pure material streams. These material streams are based on material properties. The material properties should be distinctly identifiable. Thus, although disassembly effort and visual identification are not that important, material selection is critical (Coulter et al., 1996).

6.10 DESIGN GUIDELINES FOR MATERIAL SEPARATION

We have examined the critical role played by material separation in designing products for ease of recycling. Material separation can be facilitated by adhering to the following design guidelines.

- Whether or not a complex assembly can be economically recycled depends upon the extent to which it can be separated into pure material streams.
- As we have examined in the previous section, both manual material separation and mechanical material separation have their respective advantages and applications.
- Materials should be manually separated only if significant value is retained in a part.
- Depending on whether manual separation or mechanical separation will be adopted, different design techniques should be used. This will lead to distinct designs (Coulter et al., 1996).

6.11 APPLYING MATERIAL SELECTION GUIDELINES TO DIFFERENT SCENARIOS

Most products comprise two main constituents, namely individual components and fasteners. Each of these is made of some type of material, which is our primary interest when designing product for ease of recycling. The aforementioned design rules are applied to various parts of a product in accordance with the type of material that was selected for that part. Thus, material selection plays a critical role in incorporating design features into a part. This section provides specific examples of such design decisions.

1 Component design: Modern automobiles incorporate large number of electronic items such as radios, instrument clusters, climate control units, engine control units and so on. Most of these items, by themselves, are very difficult to disassemble manually. There is one exception though. The housing for such items constitutes the largest volume of material that can be recycled. Thus, mechanical disassembly of the housing can be accomplished easily if the right material is chosen.
2 Fastener design: If a casing is made from thermoplastic, it would be advisable to use snap fits or even a few screws if the goal is to separate it manually. If, on the contrary, the same casing was to be separated mechanically, it could be welded together sonically. In this instant, the case would be shredded prior to separating the material (Coulter et al., 1996).
3 Material selection: Of the different types of plastic, thermoplastics are separated based on material density. This type of plastic is often used in automotive instrument panels. If such a panel is to be disassembled manually, it could utilize a top spin substrate composed of a blend of polycarbonate and ABS (specific gravity of about 1.1). This could be used together with a module case made from glass-filled polypropylene with the same density (Coulter et al., 1996).

   If the goal was to separate the instrument panel mechanically, it would not be possible for the above materials to be separated. Also, the polypropylene and ABS are incompatible for recycling. This would render both component unrecyclable (Coulter et al., 1996). Thus, it would be necessary to select some other material combination that is compatible for recycling.

6.12 WHAT IS REMANUFACTURING?

Remanufacturing can be defined as a process that involves restoration of old, broken products to "like new" condition by rebuilding and replacing their component parts. It involves a considerable amount of restoration. The "like new" state is not only aesthetic but also functional in nature. The process is also referred to by the more popular term: "Reman." The old, broken products are referred to as "cores."

The definition of remanufacturing is sometimes extended based on the end use of the remanufactured product or component. If new features are built into the product during remanufacture, the process is referred to as upward remanufacturing.

Succinctly, the process is defined as disassembly, cleaning, inspection, repair, replace and reassembly of the components of a part or product. The goal is to return it to "as new" condition and structure it into a new or "next generation" system (Nasr and Thurston, 2006). The distinction between upward remanufacturing and remanufacturing is quite clear.

The process of remanufacturing involves the collection of reasonably high volumes of similar products in a central place. These products are then disassembled and treated to be reused or upgraded (Sundin, 2004).

While remanufacturing has been widely practiced in the US and Europe, it is beginning to gain in importance in countries such as China. One of the reasons for this phenomenon is the fact that the number of vehicles on Chinese roads has seen a dramatic increase in recent years (Hatcher et al., 2013). Also, it has been shown that manufacturing generates more than 60% of annual nonhazardous waste (Nasr and Varel, 1996). End-of-life processes such as remanufacturing help alleviate the detrimental environmental impact of products and manufacturing processes. For instance, landfill can be reduced by requiring producers to recover used products by resorting to practices such as remanufacturing.

A very broad spectrum of products is remanufactured every year. This industrial process can be readily applied to the following disparate types of products, to name a few:

- Personal computers.
- Photocopiers.
- Cathode ray tubes.
- Industrial robots.
- Vending machines.
- Construction equipment.
- Medical equipment.
- Heavy duty engines.
- Aircraft parts and military vehicles.
- Cellular phones.
- Automobile parts.

It should be obvious from this list of products that the process of remanufacturing is almost universally relevant in our everyday life. Many companies in the automobile component remanufacturing sector often use the term "rebuilding" instead of remanufacturing. Tire manufacturers refer to themselves as "re-treaders" and companies remanufacturing laser toner cartridges refer to themselves as "rechargers" (Sundin, 2004).

Remanufacturing contributes to environmental sustainability by being less harmful to the environment than conventional manufacturing. It can also be profitable. Environmental sustainability is achieved by reducing landfill and levels of virgin material, energy as well as specialized labor used in the manufacturing process (Lund, 1984; Lund, 1996; Guide, 1999; Hormozi, 1996; McCaskey, 1994).

In order to restore a used product to "like new condition" through remanufacturing, a variety of subprocesses need to be adopted. They are delineated as follows.

1 Inspection: The used product needs to be thoroughly inspected for signs of wear, corrosion and so on.
2 Disassembly: The used product needs to be either partially or totally disassembled in order to access worn out and/or obsolete components. These components can then be repaired or replaced in order to return the product to "like new" condition. Please refer to the chapter on product disassembly for more detailed information on this topic.
3 Cleaning: The used product needs to be cleaned of grease and grime to further upgrade its condition.
4 Reprocessing: The used worn out and/or obsolete components are replaced and reprocessed. This is a critical step in the remanufacturing process. It should be borne in mind that only some components can be repaired; it is not possible for all components to be repaired. In such cases, they will have to be replaced with new components. In such cases, the remanufactured product may be sold at a lower price than its new counterpart. Even so, it is still offered with an equal warranty as a new product (Amezquita, 1996).
5 Reassembly: During this stage, the product is put back together or reassembled. This is important in order to ensure the functionality of the product as intended. Please refer to the chapter on product assembly for more detailed information on this topic.
6 Testing: No matter how well the product has been repaired, reprocessed, cleaned and reassembled, it still needs to be tested in order to verify that it does indeed function as intended and as close to a new product as possible. This is ensured by the testing phase.

6.13 COMPARING PRODUCTS THAT HAVE BEEN REMANUFACTURED, RECYCLED, RECONDITIONED AND REPAIRED

While remanufactured products are everywhere and the advantages of the process are seemingly numerous, it is still important to put the process in perspective. Remanufacturing is not the same as recycling. The two processes are substantially different. Remanufacturing is often preferred to recycling. The reason for this is because remanufacturing adds value to waste products by returning them to working order. Recycling, on the other hand, breaks down the used product to its raw material value (Ijomah et al., 2007). .Assuming that the product being remanufactured is in keeping with the production characteristics for the process, the amount of energy required to remanufacture a product is substantially less than that required for recycling (Lund, 1996).

Remanufacturing is also often confused with the term reconditioning. It should be remembered that while a remanufactured product has a warranty equal to that of a new product, a reconditioned product does not offer a similar warranty.

A remanufactured product is also substantially different from a product that has been "repaired." In the case of a repaired product, specific faults in the product have simply been repaired. Although a remanufactured product often requires more work and thus higher level of upfront energy, expense, effort and time than a reconditioned or repaired product, it results in an end product that exhibits better quality and an extended useful life (Hatcher et al., 2013).

Thus, when compared to other end-of-life processes such as recycling, reconditioning and repair, remanufacturing results in greater amount of energy savings and is considered more cost effective (Amezquita, 1995).

6.14 WHAT TYPE OF PRODUCTS CAN BE REMANUFACTURED?

We have discussed the obvious advantages of the remanufacturing process. So, is it possible to remanufacture every product when it reaches the end of its useful life? Not quite. In principle, in order to be remanufactured, a product should exhibit a high degree of durability. Technically speaking, it should exhibit the following properties:

- In order to be remanufactured, a product should be able to withstand multiple lifecycles. In other words, it should be durable enough that it can be reused, albeit in modified form. An automobile engine is an example of such a product.
- A remanufacturable product should contain high value parts. A value part is defined as a part that is worth investing in. An aircraft engine with its turbines (part of high value) is an example of such a product.
- In order for a product to be remanufactured, it is important that there is market demand for such a product. Examples of remanufactured products on the UK include automotive products, pumps and compressors and off-road equipment (Berko-Boateng, 1993).
- Efficiency of the remanufacturing process also depends to a large extent on design variables. We have already discussed the importance of such variables in chapters on design for assembly and disassembly. Design variables are defined as variables that are largely within the control of the designer. Attributes and design features such as product architecture, choice of materials, choice of fastening and joining methods and so on directly impact constituent processes such as disassembly, reprocessing, reassembly. This understanding has led to the formulation of design for remanufacturing (DfRem) guidelines. In spite of this, very few companies currently use these guidelines or actively design their products for ease of remanufacturing (Fiksel, 1996).
- In general, the following four properties are found to be the most technically relevant and important for remanufactured products and its parts (Sundin 2004):
  1. Accessibility: This property refers to the ease with parts can be accessed. Better accessibility implies greater ease of manufacturability. This is a design feature. Please refer to the chapter on design for disassembly to further understand this concept better.
  2. Ease of identification: This property refers to the ease with which parts can be identified. Greater ease of identification generally results in greater ease with which products can be remanufactured.
  3. Wear resistance: This property refers to the fact that a product or a specific component can withstand greater degrees of wear during use. As such, such products tend to have a greater longevity and are ideal candidates for remanufacturing.

TABLE 6.3
Relationship between generic remanufacturing process steps and preferred product properties

	Remanufacturing step						
Product property	**Inspection**	**Cleaning**	**Disassembly**	**Storage**	**Reprocess**	**Reassembly**	**Testing**
Ease of identification	X		X	X			X
Ease of verification	X						
Accessibility	X	X	X		X		X
Ease of handling			X	X	X	X	
Ease of separation			X		X		
Ease of securing						X	
Ease of alignment						X	
Ease of stacking				X			
Resistance to wear		X	X		X	X	

Source: Adapted from Sundin (2004).

4 Ease of handling: This property refers to the ease with which a product or component can be handled. Once again, this is a design feature and is a function of design attributes such as shape and size of the product/component, its material and so on. Please refer to the chapters on design for assembly and disassembly to gain a better understanding of these attributes. A product/component that is easy to handle is also, in many cases, easier to remanufacture. Table 6.3 illustrates the correlation between different product properties and different steps inherent in the remanufacturing process.

6.15 CONDITIONS NECESSARY FOR REMANUFACTURING TO BE PROFITABLE

Although it is possible to remanufacture products exhibiting the technical characteristics enumerated in the previous section, the profitability of the process depends on a different set of variables. Some of these variables are related to those presented in the above section (such as nature of cores), while others are different. They are described here.

- For a remanufacturing process to be profitable, it is necessary that the product contains a core, which is not consumed, discarded or is not functional (Lund, 1996).
- Remanufacturing can be profitable only if the product can be restored to its original state by using currently prevailing technology.
- It is difficult for remanufacturing to be profitable if it must rely on single-part manufacturing processes. The ability to be mass produced in a factory contributes significantly to the profitability of remanufacturing.

- After the product has been remanufactured, it is important that its value is close to the original product's market value. This is crucial for the profitability of the remanufacturing process.
- One of the key processes inherent in the remanufacturing process is the collection and acquisition of discarded and failed products. The profitability of remanufacturing depends on the extent to which such products can be acquired at relatively low cost. The acquisition cost should be low relative to the market value of the remanufactured product.
- The product design should be relatively unchanged if remanufacturing is to be profitable. Similarly, the underlying technology responsible for manufacturing should not be subject to rapid changes. If product design and/or technology changes constantly, it is quite difficult to keep pace with such changes resulting in reduced profitability.

6.16 TYPES OF REMANUFACTURERS

Given that remanufacturing is primarily a process that occurs at the end of life of a product, remanufacturing companies are variously related to the original product manufacturer. Original product manufacturers are also referred to as OEMs. Depending on their relationship with the manufacturer of the original product, remanufacturers are further subdivided into three categories. They are described in detail as follows:

- Original equipment remanufacturers (OEM): They are OEMs that remanufacture their own products. They are also referred to by the acronym OERs. Products that arrive from service centers, trade-ins from retailers and end-of-lease contracts are remanufactured by the OEM/OERs. The reason for this is because the process of remanufacturing is profitable for such companies. They can also offer a wider price range of products to their customers. Another distinct advantage that OEM/OERs possess is that they already have access to all relevant information pertaining to product design, service knowledge and logistical information such as availability of spare parts.

  In the case of OEM/OERs, the remanufacturing process could either be integrated with the original manufacturing process or it could be distinct from it. Product could either be entirely remanufactured or parts from the remanufactured products could be used in manufacturing. Examples of OEM/OERs include companies such as Caterpillar and FUJI Film. In the case of Caterpillar, several products such as engines (which the company manufactures) are also routinely remanufactured. In the case of FUJI Film, on the contrary, the company remanufactures its own single-use cameras at the same facility that it manufactures them (Sundin, 2004).
- Contracted remanufacturers: Contracted remanufacturers are companies that remanufacture products for other companies under contract. While the OEM manufactures the product, the remanufacturing is performed by a different company. After the remanufacturing has been accomplished, the remanufactured products are available for sale albeit at a lower price. As far

as the remanufacturer (independent company in this case) is concerned, the process offers several advantages. For instance, the remanufacturing operation signifies a predictable and stable revenue stream characterized by a considerably lower requirement for working capital. The OEM often assists the remanufacturer in the process by providing assistance in the form of replacement parts, tooling as well as design and testing specifications (Lund, 1983). Our-Way Inc., a refrigeration compressor remanufacturer based in Atlanta, is an example of a contracted remanufacturer.

- Independent remanufacturer: Independent remanufacturing companies are distinguished from contracted remanufacturers in that they have very little contact with the OEM. Such companies receive compensation from the previous owner or distributor to collect end-of-life products (Jakobsson, 2000). It is imperative for such companies to purchase spare parts for the products that they are remanufacturing. Independent remanufacturers often rely on an integrated operation that is characterized by purchasing cores, remanufacturing cores and marketing finished products either under its own name or under the private label of some other company. Relationship between independent remanufacturer and OEMs is mostly nonexistent and is not projected to not likely row in the future (Hammond et al., 1998). 24 Hour Toner Services, a toner cartridge remanufacturer based in Toronto, is an example of an independent remanufacturer. It should be mentioned at this juncture that a remanufacturer could contract some of its products with OEMs. For instance, Electrolux AB—a household appliance manufacturer based in Sweden—remanufactures two types of appliances. First, they remanufacture their own appliances. They also remanufacture appliances for a Danish leasing company L'Easy under contract.

It is clear from Table 6.4 that unlike manufacturing operations, used products are not necessarily owned by the company remanufacturing them (remanufacturer). To quote an example, from a 1983 investigation, 127 remanufacturers were asked, whether the "used product" was owned by them, the "product user," the OEM or by someone else. The question was asked across different market segments such as automotive, industrial and commercial. Table 6.4 depicts the responses to this question based on market segment.

It can be observed from Table 6.4 that if the remanufacturer has a contract with the OEM or if they are somehow related to the OEM, the OEM usually retains ownership of the product. Thus, money accrued as a result of building up the inventories of remanufactured parts is retained by the OEM. On the other hand, if the part has been remanufactured by a contract remanufacturer, the money retained by stored parts and work in process (WIP) is connected to the OEM. On the basis of the laws of extended producer responsibilities, the manufacturer of the product is responsible for end-of-life treatments, not the remanufacturer.

6.17 GUIDELINES FOR DFREM

DfRem can be considered to be part of a larger design for environment (DfE) approach, which has been defined by Dewberry and Goggin (1996) as an approach to design

TABLE 6.4
Remanufacturing product ownership based on market segment

Market segment	Remanufacturer (%)	Product user (%)	OEM (%)	Other (%)	Total number
Automotive • Includes automobiles, trucks, buses, motorcycles and their parts	94	4	2	0	54
Industrial • Includes all types of equipment and machinery used in manufacturing or construction.	64	25	6	5	45
Commercial • Includes all equipment used in trade or service type businesses	51	39	6	4	28
					127

Source: Sundin (2004).

wherein all environmental impacts of a product over its useful life are considered. Manufacturing companies can substantially reduce their impact on the environment using DfE strategies. The DfE approach is based on the design for X (DfX) philosophy. In order to better understand DfRem guidelines, it would be appropriate to first consider the DfX method stepwise. The different steps of DfX are presented in a bulleted format as follows (Huang, 1996):

- Collect and present facts pertaining to product and processes.
- Analyze extant relationships between products and processes.
- Gauge performance.
- Identify and analyze strengths and weaknesses and compare alternatives.
- Analyze a design from the point of view of redesigning it.
- Analyze design tradeoffs and "what if" effects.
- Implement design improvements.
- Perform iterations in the design process in order to optimize the design for the particular X under consideration.

The major DfRem considerations are described succinctly as follows:

1 Material selection: Selecting the appropriate material is at the core of any process involving product design and development. We have already discussed this at some length in chapter 1 of this book. Material characteristics play a crucial role in remanufacturing processes as well. For instance, consider the example of an engine block that is being remanufactured. Such a product could undergo several processes throughout its entire life cycle. This means it could be remanufactured several times using different processes. In order to successfully perform such operations, it is important that the block be

made from a durable material that also exhibits other characteristics such as high resistance to corrosion, resistance to wear as well as satisfactory fatigue resistance (Yang et al., 2016).

2 Type of material joining method: Material joining method implies the manner in which different materials and components are joined and connected together. This typically involves fasteners such as nuts, bolts, rivets, screws, staples, retaining lugs, adhesive joints, welds, crimps, magnets and so on. This is a very important design feature especially from the perspective of remanufacturing. This is because products are remanufactured at the end of their useful lives. The type of material used in the joining method plays an important role in deciding end-of-life options. The manner in which parts are joined together has a crucial bearing on whether the product can be reused, recycled or remanufactured. It is generally not desirable for any valuable component to inadvertently undergo damage during disassembly since it may render the part and sometimes the entire product unusable. For instance, consider two parts that have been joined using a built-in spat fit. We have already seen in the chapter on design for assembly that snap fits reduce assembly time significantly and are thus being increasingly used to join distinct parts together. However, the situation is quite different when the product is destined for remanufacturing at the end of its useful life. Although snap fit may provide fast assembly and disassembly, a failed or broken snap fit cannot be salvaged very easily and renders the part in question unfit to be reused.

3 Design of product structure: Remanufacturing of complex products is a challenging task. As we have seen before, this process involves several individual processes such as disassembly, cleaning, reassembly and so on. It is important to resort to nondestructive disassembly in order to separate components. Component design characteristics such as the number of components, design tolerance, component shape and their positioning also directly impact the efficiency of other remanufacturing processes such as cleaning, inspection, reconditioning and so on (Yang et al., 2016). For instance, a part that needs to be replaced on a frequent basis but is located deep inside the product structure and is difficult to access will have a detrimental impact on the efficiency of the remanufacturing process.

4 Surface coatings: Surface coatings are generally used when a substrate material has been selected for its bulk design characteristics. These characteristics are in contradiction to its surface design properties. Surface coatings are applied to the substrate in order to meet those requirements (Yang et al., 2016). The requirements can include properties such as wear, corrosion, aesthetic purpose or surface fatigue resistance. If a surface coating method is improperly selected, it can increase the frequency at which the product fails due to material wear or corrosion. It can also make the product remanufacturing process especially burdensome. This is because a very smooth surface coating is likely to take a substantial amount of effort to be restored to a "like new" condition. Conversely, if the coating is too coarse, it is likely to act as a dirt trap and unnecessarily complicate the cleaning process.

Table 6.5 illustrates design features that affect the ease with which a product can be remanufactured. A score of 1 implies that the design feature (such as using screws or rivets) has a low impact. A score of 4, on the other hand, points to a design feature that has a high impact.

High-level designs for remanufacturing guidelines are presented in Table 6.6. These are based on findings collected from literature and case studies. Product

TABLE 6.5
Design features that affect ease of product remanufacture

1 Low-impact	2	3	4 Very high–impact
Design features	**Problems identified**	**Impact severity**	**comments**
Type of assembly		1–4	Assembly type may hinder disassembly, an essential and initial activity that could altogether hamper remanufacturability
	Screws	1	Time consuming but does not make remanufacturing impossible
	Rivets	2	Time consuming but does not make remanufacturing impossible
	Welding	3–4	Difficult if not impossible to disassemble; depends on the type of weld
	Strong adhesive such as epoxy	4	If the adhesive is very strong, it can completely prohibit disassembly
Complexity of products		3	
	Many components	2	Many components require more resources for testing and remanufacture
	Product dimension	2	Size and weight can impede remanufacturing by obstructing access to damaged components
	Arrangement of internal components	2–3	Spatial arrangement of internal components may lead to wear due to friction between parts; also, parts, especially damaged ones may be difficult to access, thus hindering the ability to remanufacture the product: Could be caused due to ineffective communication between end user, remanufacturer, manufacturers and designers
	Coatings	2	Remanufacturing can be hindered due to ineffective and unnecessary coating, for instance, Teflon coating that flakes may result in debris that causes damage to components
Materials		4	It is not possible for nondurable materials to be remanufactured; banned materials will also hamper remanufacturing.
Design cycle		3	Resources need to be expended to stay abreast of current trends

Source: Ijomah et al. (2007).

TABLE 6.6
High-level guidelines for product remanufacturing

	Characteristics of product/design		
Process activities	**Material**	**Technique of assembly**	**Structure of product (dimensions, internal arrangement and external features)**
Product disassembly	If components are to be reused, it should be ensured that their materials are sufficiently durable to survive disassembly	Assembly methods that allow disassembly without damaging components should be used	Components should be arranged to achieve easy disassembly
	The material of the fasteners should be similar or at least compatible with that of the base material, which would limit the opportunity to damage parts during disassembly		Total number of parts should be reduced
			Complexity of disassembly should be reduced, for instance, by using standardized fasteners
			Complexity of disassembly can be reduced by using modular components, thus reducing the different types of assembly techniques
			Components should be arranged in a manner that allows for easy access and identification of separation joints
			Number of joints should be minimized
			The number of redundant parts should be reduced and/ or eliminated
			Component fits should be simplified and standardized
Cleaning components	Cleaning material that would survive cleaning processes should be used, for example, material whose melting point is higher than cleaning process temperature should be used	Only those assembly methods that allow disassembly to the point that internal components can be accessed during cleaning should be used	All parts to be cleaned should be easily accessible
	The number of material types per part should be limited		Redundant parts should be reduced or eliminated

(continued)

TABLE 6.6
(Cont.)

	Characteristics of product/design		
Process activities	**Material**	**Technique of assembly**	**Structure of product (dimensions, internal arrangement and external features)**
	Components requiring similar cleaning procedures and cleaning agents should be identified		Components should be arranged so that all can be accessed for effective cleaning
			It should be ensured that product surfaces are smooth and wear resistant
Remanufacturing components including test components	Only materials that are at least durable enough to withstand the remanufacturing process should be used	Assembly methods should be used that allow disassembly at least to the point that internal components and subsystems requiring work can be accessed	Redundant parts should be reduced or eliminated
	Materials that do not prevent upgrade and rebuilding of the product should be used	Only use assembly methods that do not prevent the product from being upgraded	Product should be structured to facilitate ease of product upgrade
	Component material should be identified	Joining methods should be used that allow disassembly at least to the point that internal components and subsystems that require it can be accessed for testing pre- and postrefurbishment.	Components should be arranged so that parts prone to damage are easily accessible
		Fault tracking devices should be incorporated	Standardized parts should be used
			Structure of the product should allow for determining component condition easily
			Product structure should allow for sequential testing: Reassembly order should be mirrored

			Disassembly level required to effectively test components should be minimized
			Test procedures should be standardized
			Component load limits, tolerances and adjustments should be clearly identified
Product assembly	The number of different materials should be limited	Components requiring similar assembly tools and techniques should be identified	Structural complexity should be reduced
		Assembly methods should not prohibit disassembly without damage to reusable components	Sequence of component assembly should be identified
		Assembly methods should simultaneously facilitate easy disassembly without damage to reusable components	Redundant parts should be reduced
		Design for assembly methods should be used: These methods should not prevent disassembly and should not damage components	Parts should be standardized
		Complexity of reassembly should be reduced by taking actions such as standardizing fasteners	Product should be structured to facilitate access to parts that have a short life and are prone to breakdown easily
			Modular structure should be adopted in such a way that obsolescence occurs with components and not the entire product

Source: Ijomah et al. (2007).

design characteristics are clearly enunciated for each type of activity inherent in remanufacturing (such as disassembly, cleaning, assembly and remanufacturing).

DfRem consideration from the perspective of material selection, joining method, product structure and surface coatings is depicted in Table 6.7.

6.18 BENEFITS OF REMANUFACTURING

As far as the OEM is concerned, remanufacturing offers the following benefits:

- Given the fact that the OEM manufactured the original product, it has complete access to the product's design content and specifications. Thus, it is also in the unique position to estimate not only the durability but also the reliability of the product. Access to the aforementioned information enables the OER to appropriately plan the remanufacturing process. Decisions pertaining to the nature of material that can be recovered from the product, how the material and product can be modified as well as the disassembly process inherent in remanufacturing are facilitated. The same can be said for decisions on the level of required maintenance.
- After having manufactured the product, the OEM sells it. As such, an established network for the distribution of the original product is easily accessible to the OEM. Since the OEM is also the OER, the same can be said for the OER. Thus, the OER also has access to the same network for the distribution of the remanufactured product. It can also access the network for the collection of discarded products. The OER is also able to build a relationship with the customer. The information obtained through this relationship can be used in the remanufacturing operation in terms of the nature of end-of-life products, the relevant schedule and quantities.
- In order to manufacture its products, the OEM needs to first establish a supplier network to enable it obtain original parts. The OER can also rely on the same network to obtain original parts for the remanufactured product. These parts would be difficult to procure from other vendors in the absence of the supplier network. It should be mentioned that such is the condition of independent remanufacturers who do not have access to a similar network. They must depend on replicas of parts and components or purchase them directly from the OEM.
- It is also possible for the OEM to leverage the knowledge gained from its supply chain network to better understand its customers. This information provides valuable insight into user patterns and can be used to evaluate residual value in discarded products.
- Information gained from its customers also enables the OEM to market its products better. One of the advantages lies in the fact that the market size can be determined more accurately. This can in turn enable the OEM/OER to remanufacture products based on demand that has been more accurately estimated. The remanufactured products can also be better marketed to a segment of the market that is more receptive to such products.

TABLE 6.7
DfRem considerations

	Material	Material Joining method	Product structure	Functional and decorative surface coating
Durability	• Corrosion resistance • Wear resistance • Fatigue resistance	• Corrosion resistance		• Wear/corrosion/fatigue resistance: Functional coating • Fingerprint/scratch resistance: Decorative coating • Adhesion
Ease of disassembly and assembly		• Nondestructive disassembly (including fastener/joint) • Nondestructive disassembly (excluding fastener/joint) • Destructive disassembly (for recycling) • Ease or reassembly	• Modularity for easy separation • Accessibility to valuable and reusable components	
Ease of cleaning	• Ease of removing impurity and deposit • Resistance to cleaning		• Avoid intricate or unnecessary concealed design form	• Ease of removing contaminants (coating removal not required) • Potential damage to the substrate (coating removal is required)
Ease of restoration and ease of upgrading	• Ease of machining • Ease of receiving additive process • Ease of receiving conditioning process • Reliability of the reconditioned part		• Accessibility to the failure prone parts • Tolerance design for multiple cycles • Modularity for replacement/ upgradability	

(*continued*)

TABLE 6.7
(Cont.)

	Material	Material Joining method	Product structure	Functional and decorative surface coating
Environmental health and safety	• Recyclability • Air emissions and waste disposal • Toxicity • Scarcity of raw material • Laws and regulations	• Compatibility with other parts • Toxicity		• Air emissions and waste disposal • Recyclability • Law and regulation
Cost	• Cost of raw material	• Labor cost • Capital cost		• Labor cost • Material and energy consumption • Capital cost
Complexity	• Number of materials	• Number of fastener/joint types • Number of fasteners/joints • Tool standardization • Accessibility to fastener/joint	• Number of parts and components • Standardization of parts and components	

Source: Yang et al. (2016).

- Since the OER is essentially the OEM, the OER can leverage the reputation of the OEM for producing high-quality products. This can go a long way in convincing customers of the reliability of the company's remanufactured products.
- For all practical purposes, the process of remanufacturing consists of reversing some part of the processes inherent in manufacturing products. As such, the infrastructure necessary for the manufacturing process is already in place. All the OER needs to do is reverse pertinent parts of the manufacturing process in order to accomplish remanufacturing. This reduces the need for additional investments. Readers will appreciate that the same cannot be said for independent remanufacturers.
- Generally, the manufacturing process involves much higher quantities than remanufacturing. As such, it allows for greater amount of investment in advanced equipment that can be used for remanufacturing as well.
- From the economic perspective, it is possible for the OEM to earn a higher profit margin due to the remanufacturing process. This is because parts recovered as a result of remanufacturing can also be used in the manufacturing process. This can provide a higher rate of return than if the parts were to be sold individually.
- From the design perspective, it would be to the designers' advantage to carefully monitor how well their designs perform during use as well as end-of-life phases. Doing so would enable them to modify their designs with the objective of making the product easier to remanufacture. This is, after all, the gist of DfRem.
- An OER generally tends to exhibit a higher worker productivity attributable to its factory methods. It also tends to make efficient use of its facilities, equipment and energy.
- Also, production volumes at such companies tended to be large enough that machines with at least a partial degree of automation could be utilized. Such machines are usually more expensive, and their higher cost can only be offset by producing in large quantities. The advantage of using such machines is that they can do away with a large degree of skilled labor.
- By its very nature, remanufacturing consists of salvaging more material. This reduces the need for virgin material and can do away with the higher cost of new parts (Lund, 1983).

Whether or not a company should remanufacture a product is always a management decision. Management decisions can be classified as strategic, tactical or operational in nature. They are described in more detail here. Examples are provided as appropriate.

- Strategic decision: Strategic decisions constitute decisions that are made to shape the long-term future of a business. From the point of view of remanufacturing, these decisions assess whether adopting the decision to remanufacture is strategically beneficial for the company. It is obvious that such a decision must be taken prior to the establishment of remanufacturing activities (Goodall et al., 2014). Remanufacturing decisions have been successfully incorporated into a business in a variety of scenarios such as:

1. The decision to remanufacture an entire product was taken in the case of single-use cameras. The remanufacturing was performed by independent third-party remanufacturers.
2. Automotive spare parts are often remanufactured in the form of aftermarket spare parts. The remanufacturing is performed by OEMs and licensed third-party remanufacturers.
3. Electronic game consoles are remanufactured under warranty (OEM or licensed third party).
4. Wind turbine gearboxes are remanufactured as a product/part service system. This means that the remanufacturing process is performed by an independent third-party remanufacturer.
5. Photocopiers are remanufactured as a product service system.

- Tactical decision: Tactical decisions are primarily medium term by nature. The objective of tactical decisions is to provide a method to implement the strategy that was picked by decision makers. From the point of view of remanufacturing, tactical decisions consider which products should be remanufactured (Goodall et al., 2014). The tactical decision is based upon specific product types and models in particular. It is important to develop general rules and heuristics at a tactical level in order to arrive at a logical decision as to which products to remanufacture. This is an important consideration because a typical remanufacturer routinely receives a range of cores for remanufacturing. These cores represent a large variety of product models and OEMs. Remanufacturers that deal with high value and low quantity products tactically assess products on a per product basis.
- Operational decision: Day-to-day decisions are characterized as operational decisions. In the context of remanufacturing, these decisions focus on evaluating individual products and components. Operational decisions made inside a remanufacturing facility are largely based on inspection operations. The main goal of an inspection operation is to weed out components and products that are not suitable to be remanufactured. We have already considered the different factors that contribute to the remanufacturability of a product. It is important to inspect products and components against such metrics in order to conserve precious resources by not resorting to unnecessary processing. Inspection decisions typically occur at different steps of the remanufacturing process. Different types and degrees of information are gained at each step. Inspections can range from virtual inspection to specific inspections. Virtual inspections rely on a pretty sophisticated infrastructure consisting of embedded sensors and conditional monitoring networks to collect and analyze information (Jun et al., 2007). Visual inspections, on the contrary, are the most widely used form of inspection. It is a quick and inexpensive form of inspection that is typically performed early in the remanufacturing process, mostly by operators who are trained to identify specific faults. Operators are trained to quickly assess the remanufacturability of products through visual inspection. Examples of specific inspection methods include those relying on metrological measurements and physical testing.

TABLE 6.8
Decision stages to assess feasibility of remanufacturing

Decision stage	Purpose	Information contained with description of product	Potential users
Strategic	Provide an early feasibility analysis of adopting remanufacturing within a business strategy	General product type	High-level management, senior and middle management
Tactical	Analyze and evaluate a product strategy for remanufacture	Specific model, product structure and bill of materials could be included	Middle management, operational management and design engineers
Operational	Evaluate a particular product for remanufacture, which can occur remotely or during inspections at a remanufacturing facility	Detailed product structure including information about product condition; other process information such as inventory levels and factory capacity may also be included	Idle management and operational management

Source: Goodall et al. (2014).

The decision stages for assessing the feasibility of remanufacturing are presented in Table 6.8.

There is a high level of uncertainty associated with the returned product cores. This introduces a significantly higher uncertainty and therefore risk in remanufacturing operations as compared to traditional manufacturing. The uncertainty is attributable to lack of information flow between early life cycle phases and the remanufacturer.

In particular, the uncertainty pertains to three factors as follows:

- The condition of product returns.
- The design and physical structure of product returns.
- The timing and quantity of product returns.

The effect of each source of uncertainty on decision making and its possible solution is presented in Table 6.8. For instance, if the condition of returned core is the source of uncertainty, Table 6.9 illustrates the impact it has on the different levels of decision making and proposes possible solutions to the problem.

The different factors affecting DfRem are encapsulated in Table 6.10.

6.19 METRIC DEVELOPMENT FOR ASSESSING REMANUFACTURABILITY

A remanufacturing metric was designed by Bras and Hammond (1996) to assess product designs for ease of remanufacturability. The metric focuses on issues that are directly affected by mechanical aspects of product design. For this reason, market issues such as core and parts availability are excluded from the metric. Core and parts

TABLE 6.9
Sources of uncertainties in remanufacturing, their effects and solutions

Source of uncertainty	Effect on decision making: Strategic	Tactical	Operational	Solutions
Condition of returned core	Added complexity in identifying the effect of long-term decision factors	Assessing the impact of uncertainties upon performance metrics including cost, quality, time and environmental impact	Measure and quantify quality of core accurately	Incorporation of multiple inspection stages Obtaining middle of life (MoL) product information
Type of returned product and design information			Determine the evaluation criteria	Important to establish links with OEM to obtain product information Need to obtain store product information obtained from experience
Quantities and timing of returns			Note inventory and production planning issues are complicated	Establish contacts with core suppliers. Offer cash back for cores.

Source: Goodall et al. (2014).

TABLE 6.10
Various factors affecting DfRem

Technical factors	Market factors	Operational factors
Ease of disassembly	Customer perceptions	Management
Ease of assembly	Product life cycle thinking	Customers
Ease of identification	Design for upgrade	Design process
Stackability		Knowledge and understanding
Accessibility		Suppliers
Resistance/durability		Cooperation
		Motivation

Source: Adapted from Hatcher et al. (2013).

availability is important to the ability to remanufacture a product (as we have already discussed at length in this chapter); however, since these are logistical issues, they are beyond the scope of the metric under consideration.

The remanufacturing metric is based on the following generic remanufacturing processes:

- Product assembly.
- Product disassembly.
- Product testing.
- Product repair.
- Product cleaning.
- Inspection of products and components.
- Product and component refurbishing.
- Component replacement.

The reader can observe that overlap of area often happens between the aforementioned processes. These areas of overlap must be eliminated so that each process can be assessed independently. This is achieved by combining or partitioning the processes into independent criteria. The list of top-level issues is refined into categories for which a set of metrics were developed. The categories are enumerated as follows:

- Cleaning.
- Damage correction: This category comprises metrics such as repair, refurbishment and replacement.
- Quality assurance: This category comprises metrics such as testing and inspection.
- Part interfacing: This category comprises metrics such as disassembly and assembly.

This categorization is presented in Table 6.11. The table depicts the composition of the remanufacturing index developed by Bras and Hammond (1996).

The structure of each of the metrics and categories is described as follows.

1 Metrics for assembly and disassembly (parts interfacing category)

The processes of assembly and disassembly are quite similar in nature. When products are being remanufactured, the disassembly sequence is often the opposite of the (re)assembly sequence. Very similar tools, techniques and fixtures are used in both processes. However, there are some design issues that are a source of complication. For instance, design features that optimize assembly do not necessarily optimize disassembly and vice versa. They could in fact serve to hinder the other process. An example is the use of snap clips in the place of screws and other threaded fasteners. The snaps can help speed up assembly times but removing them is not as easy. Thus, it is preferable to quantify assembly and disassembly separately. On the contrary, since they are so closely related, it is still desirable to consider them simultaneously when assigning weights for combination with other metrics (Bras and Hammond, 1996).

In order to compute the metrics for assembly and disassembly, it is essential to first determine the assembly and disassembly times. Next, it is important

TABLE 6.11
Composition of a remanufacturing index

Metric	Weight (%)	Index	Category	Weight	Index	Index	
Replacement (key)						Level 1	**Remanufacturing Index (Level 1 + Level 2)**
Disassembly	30		Interfacing	30			
Reassembly	70						
Testing	80		Quality assurance	5			
Inspection	20					Level 2	
Replacement	20		Damage correction	40			
Refurbishing	80						
Cleaning			Cleaning	25			

Source: Adapted from Bras and Hammond (1996).

to determine the theoretical minimum number of parts (also referred to as the number of ideal parts). In keeping with the Boothroyd and Dewhurst method, 3 seconds are allocated to each ideal part for reassembly. Sometimes, disassembly is much faster than the assembly process. In such cases, a value of only 1.5 seconds is allocated per ideal part for the process of disassembly. The metrics for assembly and disassembly efficiency are depicted in the following equations.

Disassembly efficiency = (Number of ideal parts) (1.5 seconds)/ Disassembly time.

Assembly efficiency = (Number of ideal parts) (3 seconds)/ Assembly time.

2 Metrics for testing and inspection (quality assurance category)

It is essential that remanufacturers maintain quality of their products. This is important especially in view of the lackadaisical impression that customers often seem to have about "remanufactured" products in spite of their lower cost. This concern largely is a result of the impression that remanufactured products are less capable than brand-new products. However, 100% inspection is often necessary to ensure that defective parts are not reused in the remanufactured product. This inspection practice also allows the remanufacturer to confidently issue a warranty on the product.

Inspection is not limited to visual inspection only. It is necessary that parts and/or assemblies can perform certain functions (functionally fit) within a specified set of parameters. This is where testing is important. Testing, as a rule, is more quantitative than inspection. This is because the product should perform functionally within specified limits. For instance, automotive water pumps are pressure/vacuum checked to make certain that the system is watertight (Bras and Hammond, 1996).

Inspection is defined as the qualitative inspection of parts for damage. It is usually performed during the process of disassembly or immediately after cleaning. Each part is inspected for visual damage. It is important to do so to ensure that none of the parts that goes back into the remanufactured product is visually damaged. Frequently, parts are damaged due to improper use on the part of the user, abusive environments, corrosion and so on. These are examples of sources of damage that are unexpected. In general, the inspection process examines the product beyond wear and any other damage that is usually anticipated during product design.

Whenever checking the condition of a part involves a significant investment of time and resources, such checks are like testing and are counted as such. An example would be inspecting an iron casting for cracks through the process of magnafluxing. The total number of parts that need to be inspected is designated as the total number of parts less the number of parts that are replaced. In such cases, the ideal number of inspections is the theoretical minimum number of parts that do not need to be replaced during refurbishing. The metric for inspection is given by the following equation:

Inspection efficiency = (Number of ideal inspections)/
(Number of parts–parts replaced).

Testing a product involves checking the functionality of a product. This is achieved by checking the performance of a product or subassemblies against a set of predetermined criteria. Products that are manually assembled or disassembled are also presumed to be capable of undergoing manual testing. This implies that they can be picked up and manipulated by the person in charge of doing the testing without requiring the aid of a machine. The total idealized time for testing is computed by multiplying the total number of tests by 10 seconds. The metric for testing is given by the following equation:

Testing efficiency = (Number of tests) (10 seconds)/(Testing time)

3 Metric for cleaning

Cleaning is defined as any process that involves removing foreign objects that may be present in any part. A foreign object is defined as any object that is not intended to be present in the part under consideration. The process also involves removing substances like oil and grease, which would prevent any remanufacturing processes such as painting or applying protective coatings from being performed. Parts should be designed such that markings on parts should be able to withstand cleaning. Also, surfaces to be cleaned should be smooth and resistant to wear. Part design should also enable a rational design of the cleaning line. The important thing to remember is that all deposits, impurities and other materials should be removable without causing any damage to the parts (Hundal, 2000).

It is obvious that cleaning is an important process within the overall context of remanufacturing and requires a major commitment from the remanufacturer. A major portion of the investment deals with the need to conform to environmental legislations of waste disposal requirements. Some parts such as engine heads and carburetors can also require extensive amount of cleaning. In such cases, the process can involve large amounts of time and capital expenditure. Several different subprocesses are involved in the overall process of cleaning. These can be further categorized as loose debris, dry adhered debris, oily baked debris and oily washed and dried debris (Bras and Hammond, 1996). The cleaning metric is quantified by assessing the resource requirement for each cleaning process.

Table 6.12 depicts a prioritization matrix wherein the amount of investment required for each method is compared with other methods in order to determine their relative importance. The calculated importances are rounded off wherein they constitute a set of approximations to the true relative importances. The approximations are then scaled such that the one with smallest investment is assigned a score of "1." The scaled set is used as the cleaning score for each process. The total cleaning score is computed by adding up the scores for the individual parts.

TABLE 6.12
Prioritization of different cleaning processes

	Blown	Abraded	Baked	Washed	Score	Relative importance (%)	Approximate cleaning score	Usable cleaning score
Blown	1	0.3	0.2	0.2	1.7	7	1.00	1
Abraded	3	1	0.3	0.3	4.7	18	2.69	3
Baked	5	3	1	1	10.0	38	5.77	6
Washed	5	3	1	1	10.0	38	5.77	6
					26.4	100	15.23	

Source: Bras and Hammond (1996).

The first step in creating a metric for cleaning involves development of an idealized score. One of the most basic ideal situations for cleaning constitutes blowing, brushing clean all parts of a product. This is an ideal situation since it requires the least amount of effort. A more ideal case would involve only cleaning the minimum number of parts. These parts would need to be blown free of loose debris. Thus, the metric for cleaning constitutes a comparison of the total cleaning score of each part and the ideal number of parts. This formula is given below:

Cleaning efficiency = (Number of ideal parts) (1)/(Cleaning score).

4 Metric for part refurbishing

The process of part refurbishing consists of two parts. On the one hand, it involves repair of damage to the part and application of protective and aesthetic coatings. The refurbishing process is not concerned with when the part was damaged, but only whether the damage can be rectified or undone in order to return the product back to its original capabilities. If the original capabilities cannot be restored, the part must be replaced.

Ideally, no parts should require refurbishment. This implies that in the ideal case, all parts in the product would go back into the product without having to be refurbished. As the parts that do not need to be refurbished approach the total number of parts, they approach the ideal situation. The formula for refurbishing efficiency is given below:

Refurbishing efficiency = [1 – (Number of refurbished parts)/(Total number of parts)].

5 Metric for part replacement

As we have already seen in the previous section, it is not always possible to refurbish every part. Thus, parts that cannot be refurbished or reused need to be replaced. If a very large number of such parts need to be replaced, it renders the remanufacture of the product impossible. This is because it entails a substantially large financial investment to replace all such parts. The metric for replacement is split into two: namely, a metric for replacing key parts and another for replacing nonkey parts. The key parts are construed to be essential to the functionality of the product whereas the nonkey parts are not essential.

Once again, as with part refurbishing, the ideal situation would involve not having to replace any parts at all. The metric for key part replacement is given in the following equation:

Key part replacement efficiency = [1 – (Number of key parts replaced)/(Total number of key parts)].

The remainder of the parts that are considered key parts are referred to as basic parts. Once again, ideally none of the basic parts should require replacement. The metric for basic part replacement is given below:

Basic part replacement efficiency = [1 – (Number of parts replaced– Number of key parts replaced)/(Total number of parts).

6 Structure of the combined index

The criteria from the preceding discussion are combined into one index that gauges the ability to remanufacture a product. This index satisfies the following criteria:

- The magnitude criterion: The goal of this criterion is to make sure that the remanufacturability index is not significantly larger or smaller than the metric indices. Since the metric indices have been normalized, they have a maximum value of 100%. Thus, the value of the remanufacturability index should not exceed 100% either.
- The idealization criterion: This criterion states that if all the individual indices had an ideal score of 100%, the score of the remanufacturability index should also be 100%. Given that the individual metrics were developed by normalizing them against an idealization, the index will also take on the form of comparison of actual to ideal situations. This implies that as the remanufacturability of the product approaches the ideal case, the value of the index will approach a value of 100%.
- The annihilation criterion: According to this criterion, if the value of one of the metrics approaches 0, the value of the remanufacturability metric will also approach the value of 0. This will be independent of how well the other metrics perform. This way, if a significant problem were to occur that would render the product incapable of being remanufactured, it would not be outdone by excellent performance by other metrics.
- The weighting criterion: We have already recognized the fact that the metric indices do not contribute equally to the total outcome, namely the formation of the remanufacturability index. Thus, it is necessary that each metric be weighted in keeping with the extent to which it contributes to

the overall index. This implies that a combination technique must rely on a weighting system to form the composite score.

In order to weight the different individual metrics, the authors have used the technique of inverse weighted addition. This way, each of the aforementioned four goals can be achieved.

The inverse weighted addition technique uses a nonlinear additive approach. The equation for calculating the efficiency of remanufacturability using this approach is as follows.

$$\text{Efficiency of remanufacturability} = \Sigma\ (W/\mu)^{-1}$$

Key part replacement is more important than basic part replacement as we have already discussed. For this reason, key part replacement is considered a "level 1" metric. The remainder of the metrics is considered to be "level 2" metrics. These are combined using the weighted inverted addition method. They are then combined using direct manipulation. The total combination is depicted by the following equation:

$$\text{Efficiency of remanufacturability} = (\text{Key replacement efficiency})/(\Sigma\ (W/\mu)$$

The weights for the different categories are set up using the prioritization matrix depicted in Table 6.13. The weights for the individual metrics contained in each category are presented in Table 6.11.

To sum up this discussion, the remanufacturability assessment is performed by adhering to the following steps:

1 Efficiency of each of the individual metrics is obtained by following the procedure described earlier.
2 The category indices are evaluated by combining the appropriate metrics. This is accomplished by using the weighted inverted addition.
3 The second level index is evaluated by combining the category indices using weighted inverted addition.

TABLE 6.13
Prioritization of metric categories.

	Interfacing	Damage	Quality assurance	Cleaning	Score	Exact importance (%)	Approximate importance (%)
Interfacing	1	0.2	10	5	16.2	32.5	30
Damage	5	1	10	5	21	42.1	40
Quality Assurance	0.1	0.1	0.1	0.1	1.3	2.6	5
Cleaning	0.2	0.2	1	1	11.4	22.8	25

Source: Bras and Hammond (1996).

4 The remanufacturing index is evaluated by multiplying the first level (corresponding to key part replacement) and second level (corresponding to base part replacement) and thus combining them.

6.20 CASE STUDY TO EVALUATE PRODUCT REMANUFACTURABILITY

A case study to evaluate ease of remanufacturability of an actual product has been presented in this section (Bras and Hammond, 1996). The product under consideration is a Kodak Funsaver camera. The case study consists of opinions collected from designers in the form of a questionnaire (Table 6.14) and then converting them into metrics, categories and process efficiency as described in the preceding section of this chapter.

The metric calculations are performed as follows:

1 Disassembly and assembly: Column L in Table 6.14 corresponds to the evaluation of designer responses from columns B through E. It evaluates whether each part is considered an ideal part. If any of the criteria in the columns is answered yes (Y), then the corresponding part qualifies as an ideal part and vice versa. The disassembly time is computed for each part (columns B through E in Table 6.15) and reassembly time (columns F through H). If the part is subject to corrosion, the amount of time required to disassemble that part is doubled.

2 Inspection: If an ideal part has to be replaced, it should be excluded from the number of ideal inspections, or else it is added. The total number of parts is computed by adding values in column A. The number of parts is computed from column N.

3 Testing: Column A from Table 6.14 corresponds to the total number of tests. The total testing time is computed by adding the values in column D in the same table.

4 Cleaning: The cleaning score is computed by adding the scores in column K in Table 6.14.

5 Part refurbishment: The total number of refurbished parts is ascertained by adding the value in column M of Table 6.14. This column checks columns H through K to find out whether a part needs to be replaced. The part must be refurbished if either of those columns corresponds to a Y.

6 Part replacement: We have already seen from the preceding discussion that column M checks whether a part needs to be replaced. The number of key parts is computed by summing up the values in column P. If column F has a Y value, the part is considered a key part and vice versa. Column Q checks columns P and N to determine the number of key parts replaced. If both columns show a "non-0" value for the same part, it implies that the part is a basic part that needs to be replaced.

TABLE 6.14
Design questionnaire and results for Kodak funsaver camera remanufacturability

Kodak Funsaver camera		Please answer "Y" or "N" to the following											Questionnaire results					
		Number of parts	Large relative motions?	Different material properties required?	Required to facilitate assembly or Disassembly?	Required to isolate wear?	Significant intrinsic value relative to Assembly?	Does part fatigue?	Will parts require adjustment?	If coated, can coating be reapplied?	If worn, can worn surface be replaced?	If damaged during disassembly, can damage be refurbished?	Theoretical minimum number of parts	Number of refurbished Parts	Total number of replaced Parts	Number of ideal inspections	Number of key parts	Number of key parts replaced
Part no	Part name	A	B	C	D	E	F	G	H	I	J	K	L	M	N	O	P	Q
1	Camera body	1	N	N	Y	N	Y	N	N				1	0	0	1	1	0
2	Internal aperture	1	N	N	N	N	N	N	N				0	0	0	0	0	0
3	Firing lever	1	Y	N	N	N	N	N	N				1	0	0	1	0	0
4	Spring—firing lever	1	N	Y	N	N	N	N	N				1	0	0	1	0	0
5	Cam follower	1	Y	N	N	N	N	N	N				1	0	0	1	0	0
6	Trigger catch	1	N	N	N	N	N	N	N				0	0	0	0	0	0
7	Film advance wheel	1	N	N	Y	N	N	N	N				1	0	0	1	0	0
8	Film advance cam	1	Y	N	N	N	N	N	N				1	0	0	1	0	0
9	Film winding wheel	1	N	N	Y	N	N	N	N				1	0	0	1	0	0
10	Film position wheel	1	Y	N	N	N	N	N	N				1	0	0	1	0	0
11	Top cover	1	N	N	Y	N	N	N	N				1	0	0	1	0	0

(*continued*)

TABLE 6.14
(Cont.)

Kodak Funsaver camera		Please answer "Y" or "N" to the following											Questionnaire results					
		Number of parts	Large relative motions?	Different material properties required?	Required to facilitate assembly or Disassembly?	Required to isolate wear?	Significant intrinsic value relative to Assembly?	Does part fatigue?	Will parts require adjustment?	If coated, can coating be reapplied?	If worn, can worn surface be replaced?	If damaged during disassembly, can damage be refurbished?	Theoretical minimum number of parts	Number of refurbished Parts	Total number of replaced Parts	Number of ideal inspections	Number of key parts	Number of key parts replaced
Part no	Part name	A	B	C	D	E	F	G	H	I	J	K	L	M	N	O	P	Q
12	Flash assembly	1	N	Y	N	N	Y	N	Y				1	1	0	1	1	0
13	Shutter	1	Y	N	N	N	N	N	N				1	0	0	1	0	0
14	Shutter spring	1	N	Y	N	N	N	N	N				1	0	0	1	0	0
15	External aperture	1	N	N	N	N	N	N	N				0	0	0	0	0	0
16	Lens	1	N	Y	N	N	N	N	N			N	1	0	1	0	0	0
17	Front cover	1	N	N	Y	N	N	N	N				1	0	0	1	0	0
18	Film spool	1	Y	N	N	N	N	N	N				1	0	0	1	0	0
19	Film	1	N	Y	N	N	Y	N	N				1	0	0	1	1	0
20	Back cover	1	N	N	Y	N	N	N	N				1	0	0	1	0	0
21	AA battery	1	N	Y	N	N	Y	N	N				1	0	0	1	1	0
22	Camera wrapping card stock	1	N	N	N	N	N	N	N			N	0	0	1	0	0	0

Source: Bras and Hammond (1996).

TABLE 6.15
Worksheet for the DfRem index

Kodak Funsaver camera			Disassembly				Reassembly			Cleaning		
		Number of parts	If part can corrode, it is protectively coated?	Manual removal time per part	Manual handling time per part	Disassembly time in seconds (a * f(b)* [c+d])	Manual handling time per part	Manual insertion time per part	Operating time in seconds (a*[f+g])	Cleaning code	Cleaning score per part (f(i))	Total cleaning score (a+j)
Part no.	Part name	A	B	C	D	E	F	G	H	I	J	K
1	Camera body	1		1	1			1	1	D	6	6
2	Internal aperture	1		2.1	2.1			1.7	1.7	A	1	1
3	Firing lever	1		1.8	1.8			2.2	2.2	A	1	1
4	Spring-firing lever	1		1	1			1.8	1.8	A	1	1
5	Cam follower	1		1.3	1.3			2.5	2.5	A	1	1
6	Trigger catch	1		0.8	0.8			2.7	2.7	A	1	1
7	Film advance wheel	1		0.8	0.8			1.4	1.4	A	1	1
8	Film advance cam	1		1.2	1.2			3	3	A	1	1
9	Film winding wheel	1		0.5	0.5			1.2	1.2	A	1	1
10	Film position wheel	1		0.5	0.5			1.5	1.5	A	1	1
11	Top cover	1		3.3	3.3			3.7	3.7	A	1	1
12	Flash assembly	1		2.5	2.5			6.2	6.2	A	1	1
13	Shutter	1		2.3	2.3			2	2	A	1	1
14	Shutter spring	1		2.1	2.1			4.5	4.5	A	1	1
15	External aperture	1		0.5	0.5			0.8	0.8	A	1	1
16	Lens	1		0.5	0.5			1	1		0	0

(continued)

TABLE 6.15
(Cont.)

Kodak Funsaver camera			Disassembly				Reassembly			Cleaning		
		Number of parts	If part can corrode, it is protectively coated?	Manual removal time per part	Manual handling time per part	Disassembly time in seconds (a * f(b)* [c+d])	Manual handling time per part	Manual insertion time per part	Operating time in seconds (a*[f+g])	Cleaning code	Cleaning score per part (f(i))	Total cleaning score (a+j)
Part no.	Part name	A	B	C	D	E	F	G	H	I	J	K
17	Front cover	1		2.7	2.7			2.8	2.8	A	1	1
18	Film spool	1		2.1	2.1			1.9	1.9	A	1	1
19	Film	1		1	1			15	15	A	1	1
20	Back cover	1		5.6	5.6			4.2	4.2	A	1	1
21	AA battery	1		2	2			3.8	3.8	A	1	1
22	Camera wrapping—card stock	1		4.9	4.9			10	10		0	0
					40.5				74.9			25

Source: Bras and Hammond (1996).

TABLE 6.16
Table depicting scores for different cleaning processes

Type of debris	Process	Code	Score
Loose—Powder dust	Blown/brushed	A	1
Stuck—Paint/corrosion	Abraded/buffed	B	3
Wet—Oil/dirt/debris	Baked	C	6
Wet—Oil/dirt/debris	Wash and dry	D	6

Source: Bras and Hammond (1996).

TABLE 6.17
Summary of metric computation for remanufacturing index case study

Metric	Weight (%)	Index
Replacement (key)		1.000
Disassembly	30	0.758
Reassembly	70	0.832
Testing	80	0.750
Inspection	20	0.850
Basic replacement	20	0.909
Refurbishing	80	0.955
Cleaning		0.720
Category	**Weight (%)**	**Index**
Interfacing (disassembly + reassembly)	30	0.809
Quality assurance (testing + inspection)	5	0.768
Damage correction (basic replacement + refurbishing)	40	0.945
Cleaning	25	0.720

Source: Br and Hammond (1996).

Note
Level 1 index score = 1
Level 2 index score = 0.829
Remanufacturability index = 0.829

6.21 CONCLUSION

In this chapter, we examined the basic concepts of cradle-to-grave and cradle-to-cradle manufacturing systems. We also examined the various end-of-life destinations such as reuse, recycling and remanufacturing that serve to transform a cradle-to-grave manufacturing system into a cradle-to-cradle system through conserving material. The design decisions that need to be taken early on in the product life cycle in order to enable this transformation were discussed in detail. Knowledge of these decisions in conjunction with knowledge gained earlier in the book (regarding design for assembly and design for disassembly) should enable a designer to make distinct headway in designing sustainable products.

REFERENCES

Amezquita, T. (1996). "Lean remanufacturing in the automotive industry," Master of Science Thesis, G.W. Woodruff School of Mechanical Engineering, Georgia Institute of Technology, Atlanta, Georgia.

Amezquita, T., Hammond, R. and Bras, B. (1995a). "Characterizing the remanufacturability of engineering systems," *1995 ASME Advances in Design Automation Conference*, DE-Vol. 82, Boston, MA, ASME, pp: 271–278.

Amezquita, T., Hammond, R. and Bras, B. (1995b). "Design for remanufacturing," 10th International Conference on Engineering Design (ICED 95), Praha, Czech Republic, Heurista, Zurich, Switzerland, pp: 1060–1065.

Berko-Boateng, V.J., Azar, J., De Jong, E. and Yander, G.A. (1993). "Asset recycle management—A total approach to product design for the environment," International Symposium on Electronics and the Environment, Arlington, VA, IEEE, pp: 19–31.

Bras, B. and Hammond, R. (1996). "Towards design for remanufacturing-metric for assessing remanufacturability," Proceedings of the 1st International Workshop on Reuse. Eindhoven, The Netherlands. July 19, 1996.

Coulter, S., Bras, B., Winslow, G. and Yester, S. (1996). "Designing for material separation: Lessons from automotive recycling," ASME Design Engineering Technical Conferences and Computers in Engineering. Irvine, California. August 18–22.

Dewberry, E. and Goggin, P. (1996). "Spaceship ecodesign," *Co-Design: Interdisciplinary Journal of Design and Contextual Studies*, Vols. 5–6, Nos. 1–3, pp 12–17.

Fiksel, J. (1996). "Towards sustainable development," *Design for Environment—Creating Eco-Efficient Products and Processes*. McGraw-Hill, New York, pp: 23–33.

Goodall, P., Rosamond, E. and Harding, J. (2014). "A review of the state of the art in tools and techniques used to evaluate remanufacturing feasibility," *Journal of Cleaner Production*. Vol: 81, pp: 1–15.

Guide, V.D. (1999). "Remanufacturing production planning and control: U.S industry best practice and research issues," Second International Working Paper on Re-use, Eindhoven, The Netherlands, pp 115–128.

Hammond, R., Amezquita, T. and Bras, B. (1998). "Issues in automotive parts remanufacturing industry: discussion of results from surveys performed among remanufacturers," *Journal of Engineering Design and Automation, Special Issue on Environmentally Conscious Design and Manufacturing*, Vol: 4. No: 1, pp 27–46.

Hata, T., Kato, S., Sakamoto, H., Kimura, F. and Suzuki, H. (2000). "Product life cycle simulation with quality model," Proc.7th CIRP Int. Seminar on Life Cycle Engineering, Tokyo, 60–67.

Hatcher, G.D., Ijomah, W.L. and Windmill, J.F.C. (2013). "Integrating design for remanufacture into the design process: The operational factors," *Journal of Cleaner Production*, Vol: 39, pp: 200–208.

Hormozi, A. (1996). "Remanufacturing and its consumer, economic and environmental benefits" APEX Remanufacturing Symposium, USA, pp. 5–7 May 20–22.

Huang, G.Q. (1996). *Design for X: Concurrent Engineering Imperatives*. Springer Dordecht, Chapman and Hall.

Hundal, M. (2000). "Design for recycling and remanufacturing," International Design Conference-Design, Dubrovnik. May 23–26, 2000.

Ijomah, W.L., McMahon, C.A., Hammond, G.P. and Newman, S.T. (2007). "Development of robust design for remanufacturing guidelines to further the aims of sustainable development," *International Journal of Production Research*, Vol: 45, No: 18–19, pp: 4513–4536.

Jakobsson, N. (2000). "Emerging product strategies-selling service of remanufactured products," The International Institute for Industrial Environmental Economics. Lund University, Lund, Sweden.

Jun, H.-B., Kiritsis, D. and Xirouchakis, P. (2007). "Research issues on closed-loop PLM," *Computers in Industry*, Vol: 58, pp: 855–868.

Kimura, F., Hata, T. and Suzuki, H. (1998). "Product quality evaluation based on behavior simulation of used products." *CIRP Annals*, Vol: 47, No: 1, pp: 119–122.

Kimura, F., Kato, S., Hata, T. and Masuda, T. (2001). "Product modularization for parts reuse in inverse manufacturing," *CIRP Annals*, Vol: 50, No: 1, pp: 89–92.

Kriwet, A., Zussman, E. and Seliger, G. (1995). "Systematic integration of design for recycling into product design," *International Journal of Production Economics*, Vol: 38, pp: 15–22.

Kunststofindustrie, F.N.R.e. (2015). Caseguide Ontwerpen met Kunststof Recyclaat.

Lund, R. (1983). "Remanufacturing: The experience of the United States and implications for developing countries," CPA/83-17. The World Bank, Washington, DC.

Lund, R.T. (1984). "Remanufacturing: The experience of the U.S.A. and implications for the developing countries," World Bank Technical Paper No. 3.

Lund, R. (1996). *The Remanufacturing Industry: The Hidden Giant.* Boston University. Boston, MA.

McCaskey, D. (1994). "Anatomy of adaptable manufacturing in the remanufacturing environment," APICS Remanufacturing Seminar Proceedings, USA, pp 42–45.

Miller, T. and Elgård, P. (1998). "Defining modules, modularity and modularization," Proceedings of the 13th IPS Research Seminar, Fuglsoe, Denmark.

Nasr, N. and Thurston, M. (2006). "Remanufacturing: A key enabler to sustainable product systems," Proceedings of LCE. Leuven, May 31–June 2, 2006.

Nasr, N. and Varel, E. (1996). "Lifecycle analysis and costing in an environmentally conscious manufacturing environment," APICS Remanufacturing Symposium Proceedings, USA, pp. 44–47.

Ragaert, K., Delva, L. and Geem, K. (2017). "Mechanical and chemical recycling of solid plastic waste," *Waste Management*, Vol: 69, pp: 24–58.

Sundin, E. (2004). "Product and process design for successful remanufacturing," Dissertation. Linkoping Studies in Science and Technology.

Tanaka, H. (2000). "Fully automatic inverse factory for re-using single-use cameras," Proc.7th CIRP Int. Seminar on Life Cycle Engineering, Tokyo, 256–257.

Yang, S., Ong, S. and Nee, A. (2016). "A decision support tool for product design for remanufacturing," 13th Global Conference on Sustainable Manufacturing-Decoupling Growth from Resource Use. Procedia CIRP. Vol: 40, pp: 144–149.

7 Costing for Sustainable Product Design

7.1 INTRODUCTION

In order to be truly sustainable, a product needs to address all three pillars of sustainability, namely the pillars of environmental conservation, social sustainability and economic sustainability. We have examined the role of the first two pillars in the process of designing sustainable products. The concept of designing products for ease of assembly and disassembly was discussed in detail. In addition, we have also analyzed how products can also be designed to facilitate end-of-life (EoL) objectives such as ease of reuse, recycling and remanufacturing. Furthermore, the book also presented a brief overview of environmental legislations. Steps being undertaken by key manufacturers to address the problem of waste and waste management were discussed as well.

The third pillar of sustainability, namely the economic pillar, will be dealt with in this chapter. The main objective here is to present basic economic concepts such as different types of cost that need to be borne by the manufacturer in the design of sustainable products. Examples include direct cost and indirect cost. Cost can also be classified as material cost and labor cost. It is important to remember that a sustainably designed product is likely to have a different cost structure than a product that is destined for the landfill. It is critical that the designer fully understand and account for such a cost structure. To that end, this chapter will present mathematical models which enable designers to consider the cost of designing product for various end-of-life options early during the design phase. First, we will present some basic definitions pertaining to the types of cost that are typically incurred by any manufacturer.

7.2 BASIC DEFINITIONS AND TYPES OF COST

Direct material cost: Direct cost can be defined as the amount of expenditure that needs to be incurred in purchasing the right material for the product. Direct material cost plays an important role in determining total product cost. Its role has evolved over time. In the early part of the twentieth century, a major chunk of the total product cost was attributed to direct material. This is no longer the case.

Direct labor cost: Direct labor cost can be defined as the amount of money that is paid to workers who are directly responsible for manufacturing and assembling different parts of the product. Thus, wages paid to line workers can be classified as direct labor cost. However, wages paid to engineers do not qualify as direct labor.

Indirect cost: Certain costs need to be incurred by the manufacturer of any manufacturing enterprise. It is not always possible to clearly identify and charge these costs to a particular product. Such costs are also referred to as overhead expenses.

Thus, indirect cost is the expenditure that is incurred during the production or manufacturing process that cannot be classified as any type of direct expense such as direct material cost, direct labor cost or any other direct expenses (such as engineering expenses). Thus, overhead expenses can also be termed as indirect expenses. Examples of overhead costs include administrative costs, secretarial costs, supervisor salary and so on. Indirect costs can be further subclassified as follows:

7.2.1 Types of Indirect Costs

- Indirect material costs

Regardless of its function, every product is composed of some material. Whether that material is plastic, glass, composite or some metal is not important. This material, regardless of type, is termed as direct material.

Material that is not part of the final product is defined as indirect material. This type of material is typically consumed during any manufacturing process. Examples of indirect material costs include cost of grinding wheels, cutting fluids in machining, lubricants for forming, cleaning materials, cost of lubricants such as oils, grease and other lubricants. Expenses incurred in obtaining the aforementioned products can be classified as indirect material cost. Note that although grinding wheels and lubricants, for instance, are an essential constituent of the manufacturing process, they are not a component of the finished product itself.

- Indirect labor costs

We have already defined direct labor. Indirect labor is defined as labor that plays no direct role in the actual manufacturing process itself. However, to ensure smooth functioning of the manufacturing operation, it is necessary to elicit the service of certain personnel. Salaries paid to supervisors, repair and maintenance staff, care operators, janitorial staff, foremen and miscellaneous factory personnel are examples of indirect labor costs. Although administrative staff, such as secretaries, are not directly responsible for manufacturing the products, they are nevertheless essential to the overall functioning of the enterprise. For this reason, wages paid to such staff are also classified as indirect labor cost. Unlike foremen and maintenance staff, administrative staff may not necessarily be present on the site of actual production.

- Indirect engineering costs

When a product is manufactured, there are certain expenses that need to be incurred that cannot be classified as either direct or indirect material expenses. Neither can they be classified as direct or indirect labor expenses. The cost to procure computer supplies, computer software, and any other equipment that is necessary to ensure successful design, development and production of the finished product are examples of such costs. They are typically defined as indirect engineering costs.

We have familiarized ourselves with the basic definitions of different types of costs that are incurred during a manufacturing process. The next section will present some mathematical models that can enable the designer to estimate the product development cost for different end-of-life options such as reuse, recycling, remanufacturing and refurbishing.

7.3 MATHEMATICAL MODELS TO PREDICT DEVELOPMENT COSTS FOR VARIOUS END-OF-LIFE OPTIONS

Before we delve into cost estimation for various end-of-life options, we need to first understand the critical importance of the disassembly process itself. The concept of disassembly was dealt with thoroughly in chapter 3 of this book. It was also addressed in detail in chapter 6, when we dealt with design for reuse, recycling and remanufacturing. Readers will recall that products need to be disassembled to varying degrees before they can be reused, recycled, remanufactured or refurbished. Therefore, it is critical to first estimate disassembly costs.

As we have explained earlier, disassembling a product allows components and materials to be broken down into individual parts. There are two types of disassembly processes: Nondestructive disassembly and destructive disassembly. The disassembly process helps achieve the following objectives in industry (Das et al., 2000):

- Retrieval of reusable and valuable vehicle parts.
- Removal of hazardous and toxic materials.
- Remanufacturing of components and products in order to elongate their lifespans.
- Breaking down individual components.

The following parameters affect the process of product disassembly (Bogue, 2007):

- Complexity of components.
- Types of fasteners used.
- Degree of fragility of the component.
- Wear resistance.
- Ease with which the component can be identified and handled.

An additional parameter was used to quantify the disassembly process, namely the disassembly depth factor (Cheung et al., 2015). This parameter measures the level of difficulty involved in disassembling a product. This is an important consideration which needs to be taken into account when determining disassembly cost.

The disassembly depth factor can be determined as follows:

f = (Number of assemblies to disassemble)/Total number of assemblies.

The following assumptions are made in the above calculation:

- It is possible to disassemble each assembly in the same amount of time.
- Upon being disassembled, each assembly is divided into fundamental components and low-level assemblies.
- The disassembly time for the entire assembly is equal to the time for separating individual components.

Direct labor cost for disassembly can also be computed by multiplying the disassembly time and direct labor rate. Consider a situation where it takes 10 minutes to disassemble a product and the direct labor rate is $20 per hour.

In this case, the direct labor cost for the disassembly operation would be $20 divided by 6. This equates to $3.33.

Recall that we have presented at least two case studies in chapter 3 (design for disassembly) that demonstrate computation of disassembly time. A case study was also included in chapter 4 that dealt with computation of assembly time.

Next, we present models for computing recycling cost, remanufacturing cost and refurbishing cost.

7.3.1 Recycling Cost

According to Yan et al (1995), the process of recycling consists of recovering used components so that they can be used as raw materials. We have already dealt with the concept of recycling in detail in chapter 6. The following costs need to be considered for recycling:

- Disassembly cost.
- Cleaning cost.
- Recycling cost. This includes costs for material processing, manufacturing, packaging and assembly.

Recycling cost is computed using Dantec's approach (2005) as follows:

$$\text{Recycling cost} = \Sigma\, (RV_i - RC_i)$$

Where $RV_i = \Sigma\, ((\text{Parts}_i \times MV_m - OC_i) \times W_i)$

$$RC_i = \Sigma\, (RT_i \times f \times L)$$

Where:

$Parts_i$: The number of parts of type "I."
W_i: The weight of parts of type "I" in lbs.
MV_m: The mass of material value of parts ($/lb). The original model states the value in £/kg.
RT_i: The amount of time necessary to remove one type "i" part. This is expressed in hours.
L: Hourly wage, expressed in $/hour.
OC_i: This is the opportunity cost. It corresponds to the revenue earned by the dismantler by selling the parts to the shredder (Dantec, 2005). It is expressed as $/lb.
f: Disassembly depth factor.

7.3.2 Remanufacturing Cost

According to Ijomah et al. (2007a, 2007b), remanufacturing is defined as a process of returning a used product to at least the performance specification required by the original equipment manufacturer (OEM) and giving the resulting product a warranty that is at least equivalent to that of a newly manufactured product or equivalent.

The following costs need to be considered for remanufacturing:

- Disassembly costs.
- Cleaning costs.
- Redesign costs.
- Remanufacturing costs.

According to Hatcher et al. (2013), decisions that are made during the design stage have a capacity to greatly influence the efficiency and effectiveness of the remanufacturing process. Remanufacturing cost can be computed by means of the following equation (Shu et al., 1999);

$$C_{RM} = ((T_d + T_a) \times L \times f) + (P_f \times C_f) + ((P_{pd} + (P_f \times P_{pe}) - (P_{pd} \times P_f \times P_c)) \times C_p)$$

Where:

C_{RM}: Remanufacturing cost.
T_d: Disassembly time, expressed in hours.
T_a: Assembly time, expressed in hours.
L: Labor rate, expressed in $/hour.
f: Disassembly depth factor.
P_f: Probability of fastener failure during the disassembly and assembly processes.
C_f: Cost of fastener failure, expressed in $.
P_{pd}: Probability of part failure during the disassembly and assembly processes.
P_{pe}: Probability of part failure in fastening Method extraction.
C_p: Cost of part failure, expressed in $.

7.3.3 Refurbishing Cost

Refurbishing is an important constituent of the overall manufacturing process. The goal of any refurbishment activity is to ensure that components that are reusable are as capable as new components. Irrespective of whether the component under consideration has failed, it will still be prioritized for refurbishment. A relatively minor refurbishment cost, such as cleaning cost, will be incurred if a component is in good condition at the end of its useful life. However, if the component under consideration has failed, it will need to undergo a greater degree of refurbishment such as repair or replacement. In such cases, a correspondingly high cost will need to be incurred as well.

Refurbishment cost is determined by using the following equation (Zhou et al., 1999):

$$C_{Refurbish} = \Sigma(\dot{C}_{good}^{e^{-\lambda t}} + Cfailure^{\left(1-e^{-\lambda t}\right)}) + f \times L \times \left(T_d + T_a\right).$$

Where:

- C_{Good}: Cost of refurbishing a part/component that is in good condition, expressed in \$.
- $C_{failure}$: Cost to refurbish a part/component that has failed.
- t: Expected lifetime of the product (this is usually expressed in millions).
- T_d: Disassembly time, expressed in hours.
- T_a: Assembly time, expressed in hours.
- L: Labor rate, expressed in \$/hour.
- f: Disassembly depth factor.
- □: Rate of degradation expressed as failures per million hours.

7.4 CONCLUSION

The third pillar of sustainability, namely the economic pillar, was addressed in this chapter. More specifically, the chapter presented some mathematical models that would enable the designer to estimate costs for various end-of-life options such as recycling, remanufacturing and refurbishing. It is important to note that these models enable the designer to estimate said costs early during the design stage of the product. This is an important consideration. If a particular design option is known to be cost prohibitive for the company before it is ever incorporated into the design, then no matter how environmentally sustainable it might be, the company can still weigh its options. Different options can be weighed such that a better option that is more economically sustainable can be used. The basic idea of sustainable product design is to incorporate mutually acceptable design tradeoffs in the final product design. The objective of the tradeoffs is to design a product that addresses all three pillars of sustainability, not just the environmental pillar or economic pillar or the social pillar in isolation.

REFERENCES

Bogue, R. (2007). "Design for disassembly: A critical twenty first century discipline," *Assembly Automation*, Vol: 27, No: 4, pp: 285–289.

Chenug, W.M. and Pachisia, V. (2015). "Facilitating wastepaper recycling and repurposing via cost modeling of machine failure, labor availability and waste quantity," *Resource, Conservation, Recycling*, Vol: 101, pp: 34–41.

Cheung, W.M., Griffin, P., Newnes, L., Mileham, A., Marsh, R. and Lanham, J. (2015). "Towards cleaner production: A roadmap for predicting product end of life costs at early design concept," *Journal of Cleaner Production*, Vol: 87, pp: 431–441.

Dantec, D. (2005). "Analysis of the cost of recycling compliance for the automobile industry," Thesis for a Master of Science in Technology and Policy. Massachusetts Institute of Technology, MIT, USA.

Das, S.K., Yedlarajjah, P. and Narendra, R. (2000). "An approach for estimating the end of life product disassembly and cost," *International Journal of Production Research*, Vol: 38, No. 3, pp: 657–673.

Ijomah, W.L., McMahon, C.A., Hammond, G.P. and Newman, S.T. (2007a). "Development of design for remanufacturing guidelines to support sustainable development," *Robotics and Computer Integrated Manufacturing*, Vol: 23, No: 6, pp: 712–719.

Ijomah, W.L., McMahon, C.A., Hammond, G.P. and Newman, S.T. (2007b). "Development of robust design for remanufacturing guidelines to further the aims of sustainable development," *International Journal of production Research*, Vol: 45, No: 18–19, pp: 4513–4536.

Shu, L. and Flowers, W. (1999). "Applications of a design for remanufacture framework to the selection of product lifecycle fastening and joining methods," *Robotics and Computer Integrated Manufacturing*, Vol: 15, No: 3, pp: 179–190.

Yan, X. and Gu, P. (1995). "Assembly/disassembly sequence planning for life cycle cost estimation," *Manufacturing Science and Engineering*, ASME, MED-2(2)/Mh-3, No: 2, pp: 935–956.

Zhou, M., Caudill, R., Sebastian, D. and Zhang, B. (1999). "Multi lifecycle product recovery for electronic products," *Journal of Electronic Manufacturing*, Vol: 9, No: 1, pp: 1–15.